Modern Innovations in WEED MANAGEMENT

Dr. Ujagar Singh Walia
A Life in Agronomy and Weed Science

Dr. Ujagar Singh Walia is a celebrated agronomist with over 50 years of expertise in weed management, agricultural research, and teaching. A distinguished alumnus of Punjab Agricultural University (PAU) Ludhiana, he completed his Master's in Agronomy (specializing in Weed Science) and later earned his PhD, establishing himself as a leading authority in the field.

Throughout his long and illustrious career at PAU Ludhiana, Dr. Walia served in various capacities–culminating in his role as Senior Agronomist and Head of the Department of Agronomy–focusing primarily on weed control. His research, teaching, and extension work have significantly advanced sustainable crop production and innovative weed management practices. Dr. Walia has authored 176 research papers, 158 popular articles for farmers, and 8 seminal books, including titles like *Weed Management, Identification of Weeds* and *Modern Techniques for Crop Management etc.* His work on herbicide recommendations–85 in total–has been instrumental for the agricultural community in Punjab.

An inspiring mentor, Dr. Walia has guided over 34 M.Sc. students and 12 Ph.D. scholars, nurturing the next generation of agricultural scientists. His contributions have earned him several prestigious awards, such as the ISWS Special Recognition Award (2022-23), a Gold Medal from the Indian Society of Weed Science in 2006, and the FAI Silver Jubilee Award for outstanding doctoral research in 1984.

Beyond his academic achievements, Dr. Walia has held key leadership roles, serving as Joint Secretary and Vice President of the International Society of Weed Science (ISWS). He has also participated as a visiting scientist in international projects on herbicide resistance and sustainable farming systems, collaborating with research teams in Australia and the Philippines.

Dr. Ujagar Singh Walia's legacy is defined not only by his extensive publications and research projects but also by his lifelong commitment to improving agricultural practices and empowering farmers. His work remains a beacon of innovation and dedication in the field of weed science, inspiring countless professionals and students around the world.

Modern Innovations in WEED MANAGEMENT

U.S. Walia

Sr. Agronomist Cum Head (Retd.)
Department of Agronomy
Punjab Agricultural University
Ludhiana (Punjab)

2026

Daya Publishing House®
A Division of
Astral International Pvt. Ltd.
New Delhi – 110 002

ISBN 978-93-5919-863-7

Published by : **Daya Publishing House®**
A Division of
Astral International Pvt. Ltd.
– ISO 9001:2015 Certified Company –
4736/23, Ground Floor, Ansari Road
Darya Ganj, New Delhi-110 002
Ph. 011-43549197, 8130496929
e-mail: info@astralint.com
Website: www.astralint.com

Printed at : **Replika Press Pvt Ltd**

Preface

Weeds have challenged agriculture since the earliest days of farming, competing relentlessly with crops for sunlight, water, nutrients, and space. Over time, farmers and scientists have devised a wide range of strategies—manual, cultural, mechanical, chemical, and biological—to combat these persistent adversaries. Yet, as agriculture continues to evolve under the pressures of population growth, climate change, herbicide resistance, and environmental sustainability, the need for modern, innovative approaches to weed management has never been greater.

This book, *Modern Innovations in Weed Management*, is an attempt to bridge traditional wisdom with the latest technological advancements shaping the future of agriculture. It highlights the remarkable journey of weed control—from labor-intensive practices to precision-driven systems powered by robotics, artificial intelligence, drones, sensors, genetic engineering, and nanotechnology. Alongside these technological frontiers, the book also emphasizes sustainable, integrated approaches that balance productivity with ecological responsibility.

The aim of this work is to provide researchers, students, agronomists, extension specialists, and farmers with a comprehensive resource that not only explains current methods but also envisions the path forward. Each chapter is designed to present both the science and practice of weed management, with attention to practical applications, economic considerations, and environmental impacts. By examining challenges such as herbicide resistance, regulatory frameworks, and climate variability, the book encourages readers to adopt adaptive and forward-looking strategies.

Agriculture today stands at the intersection of tradition and innovation. Just as weeds adapt and evolve, so must our methods of managing them. It is my hope that the insights presented here will inspire greater understanding, encourage responsible adoption of technology, and support the shared goal of ensuring food security while safeguarding the environment.

I express my gratitude to colleagues, students, and farmers whose questions and experiences have continuously shaped my research and teaching. Their contributions reaffirm the principle that agricultural progress is most successful when knowledge flows between laboratories, classrooms, and fields.

This book is dedicated to the continuing journey of discovery and innovation in weed science—an endeavor essential to the prosperity of agriculture and the well-being of humanity.

Dr. U.S. Walia

Sr. Agronomist & Head (Retd.),
Department of Agronomy, PAU Ludhiana

Contents

Chapter 1

The Shift from Traditional to Modern Weed Management Practices

Weed management has evolved significantly over the past decades, transitioning from traditional, labor-intensive methods to advanced, science-driven approaches. This shift has been driven by the need for higher agricultural productivity, environmental sustainability, and labor efficiency.

Traditional Weed Management Practices

- ☆ **Preventive Method**: Use clean and bold seeds which should be free from weed seeds. Clean agricultural machinery while moving from one to another field.
- ☆ **Manual Weeding**: Hand-pulling and hoeing were the primary methods, requiring significant labor and time. These methods are still prevalent in smallholder farming systems but are inefficient for large-scale operations.
- ☆ **Cultural Practices**: Techniques like crop rotation, intercropping, selection of quick growing crops/varieties, placement of fertilizers, increased seed rate and adjusting planting dates were employed to suppress weed growth. While effective to some extent, these methods lacked precision and scalability.
- ☆ **Mechanical Control**: Animal-drawn implements and rudimentary tools were used, offering limited weed control and often causing soil disturbance. Inter-row weeding can also be done with tractor operated weeders especially in wider row crops.

Modern Weed Management Practices

- **Herbicides**: The introduction of synthetic herbicides revolutionized weed control by providing selective and non-selective options. Glyphosate, for instance, became a cornerstone for managing broad-spectrum weeds (Duke and Powles, 2008).
- **Integrated Weed Management (IWM)**: Modern systems combine chemical, biological, and cultural methods to minimize herbicide resistance and environmental impact.
- **Precision Agriculture**: Technologies like GPS-guided sprayers and drones enable site-specific weed control, reducing chemical usage and improving efficiency.
- **Biological Control**: The use of natural enemies, such as insects or pathogens, to suppress weed populations has gained traction, especially in organic farming systems.

The transition from traditional to modern practices reflects advancements in agricultural science and technology, addressing the challenges of weed resistance, environmental degradation, and labor shortages.

A. Challenges in Current Weed Management

Weed management is a critical component of modern agriculture, as weeds compete with crops for essential resources such as light, water, and nutrients, ultimately reducing crop yields and quality. Despite advancements in weed control technologies, several challenges persist in achieving sustainable and effective weed management. These challenges are multifaceted, encompassing biological, technological, economic, and regulatory aspects. This section explores the key challenges faced in current weed management practices.

1. Herbicide Resistance

One of the most pressing challenges in weed management is the widespread development of herbicide resistance. Over-reliance on herbicides with similar modes of action has led to the evolution of resistant weed biotypes.

- **Mechanisms of Resistance**: Weeds develop resistance through mechanisms such as target-site mutations, enhanced metabolism of herbicides, or reduced herbicide uptake and translocation. For example, resistance to glyphosate, a widely used herbicide, has been reported in species like *Amaranthus palmeri* (Palmer amaranth) and *Lolium rigidum* (rigid ryegrass) (Heap, 2023). *Phalaris minor* developed resistance to isoptounon in Punjab (Walia *et al.*, 1997).
- **Global Impact**: As of 2023, herbicide resistance has been documented in 514 unique weed species across 94 crops in 71 countries (Heap, 2023). This resistance reduces the efficacy of herbicides, leading to increased application rates, higher costs, and environmental concerns.

2. Limited Development of New Herbicides

The development of new herbicide molecules has slowed significantly over the past few decades. This is due to a combination of high research and development (R&D) costs, stringent regulatory requirements, and the technical complexity of discovering novel modes of action.

- ☆ **Economic Barriers**: The cost of bringing a new herbicide to market is estimated to exceed $250 million, with a development timeline of 10–15 years (Duke, 2012). These costs deter investment in herbicide innovation.
- ☆ **Lack of Novel Modes of Action**: Most herbicides developed in recent years target the same biochemical pathways as older products, limiting their utility in managing resistant weed populations.

3. Environmental and Regulatory Constraints

Increasing awareness of environmental sustainability has led to stricter regulations on herbicide use, further complicating weed management strategies.

- ☆ **Regulatory Restrictions**: Many herbicides, such as atrazine and paraquat, have been banned or restricted in certain regions due to concerns about their environmental persistence, toxicity, and potential to contaminate water sources.
- ☆ **Environmental Concerns**: Excessive herbicide use can lead to soil degradation, non-target species damage, and biodiversity loss. These concerns necessitate the adoption of integrated weed management (IWM) practices, which are often more labor-intensive and knowledge-dependent.

4. Climate Change and Its Impacts

Climate change is altering weed dynamics, making weed management more unpredictable and challenging.

- ☆ **Shifts in Weed Populations**: Rising temperatures and changes in precipitation patterns are enabling the proliferation of invasive weed species in new regions. For instance, *Parthenium hysterophorus* (parthenium weed) has expanded its range due to warmer climates. Also with the introduction of Mexican high yielding wheat varieties invited *Phalaris minor* as major weed of wheat which had replaced all traditional weed flora of desi wheat varieties.
- ☆ **Extended Growing Seasons**: Longer growing seasons in temperate regions allow weeds to produce multiple generations, increasing their population density and seed bank persistence.

- **Herbicide Efficacy**: Changes in temperature and humidity can affect herbicide absorption, translocation, and efficacy, requiring adjustments in application timing and rates.

5. Labor Shortages and Rising Costs

The shift away from manual weeding due to labor shortages and rising wages has increased reliance on chemical and mechanical weed control methods.

- **Mechanization Limitations**: While mechanical weeding tools are effective in certain scenarios, they are often unsuitable for smallholder farms or fields with irregular topography.
- **Economic Pressures**: The rising cost of herbicides, coupled with the need for multiple applications to combat resistant weeds, places a financial burden on farmers, particularly in developing countries.

6. Knowledge Gaps and Adoption Barriers

The successful implementation of advanced weed management strategies, such as integrated weed management (IWM) or precision agriculture, requires significant knowledge and skill.

- **Limited Awareness**: Many farmers lack awareness of IWM principles, such as crop rotation, date of sowing crop density, cover cropping, and the use of biological control agents.
- **Technology Access**: Precision agriculture tools, such as weed-detecting drones and robotic weeders, are often expensive and inaccessible to smallholder farmers.
- **Training Needs**: Effective weed management requires ongoing education and training for farmers, extension workers, and agronomists to stay updated on the latest research and technologies.

7. Weed Seed Bank Persistence

The quantum and longevity of weed seeds in the soil seed bank poses a significant challenge to long-term weed management.

- **Seed Dormancy**: Many weed species produce seeds with dormancy mechanisms that allow them to remain viable in the soil for years or even decades. For example, seeds of *Chenopodium album* (common lambsquarters) can persist for up to 40 years under favorable conditions.
- **Management Implications**: Reducing the weed seed bank requires consistent and sustained efforts, such as timely weed removal before seed set and the use of effective weed control strategies to suppress weed emergence and their growth.

B. The Shift from Traditional to Modern Practices

The evolution of weed management practices has been a cornerstone of agricultural progress. Historically, farmers relied on traditional methods to control weeds, which were often labor-intensive, time-consuming, and less effective. However, with advancements in agricultural science and technology, modern weed management practices have emerged as more efficient, sustainable, and precise solutions. This section explores the transition from traditional to modern weed control methods, highlighting the driving factors, key differences, and the role of innovation in shaping contemporary practices.

Traditional Weed Management Practices

Traditional weed management methods have been practiced for centuries, often relying on manual labor, cultural practices, and rudimentary tools. While these methods were effective to some extent, they were limited by their inefficiency and inability to address large-scale weed infestations.

Key Characteristics of Traditional Practices

- ☆ **Manual Weeding**: Farmers primarily relied on hand-pulling or hoeing to remove weeds. While this method was precise, it was labor-intensive and impractical for large farms.
- ☆ **Cultural Practices**: Techniques such as crop rotation, intercropping, and timely planting were employed to suppress weed growth. These methods aimed to create unfavorable conditions for weeds but were not always sufficient to control aggressive species.
- ☆ **Mechanical Control**: Basic tools like plows, harrows, and weed hooks were used to disrupt weed growth. These tools required significant human or animal effort and were less effective in controlling deeply rooted or perennial weeds.
- ☆ **Burning and Flooding**: In some regions, farmers used fire or water to eliminate weeds. While effective in specific contexts, these methods often had negative environmental impacts, such as soil degradation or water wastage.

Challenges of Traditional Methods

- ☆ High labor requirements, making them unsuitable for large-scale farming.
- ☆ Limited effectiveness against invasive or herbicide-resistant weed species.
- ☆ Environmental concerns, such as soil erosion caused by excessive tillage.
- ☆ Lack of precision, often leading to damage to crops alongside weeds.

- Weeds have tendency to germinate in flushes, so these methods fail to control weeds

Drivers of the Shift to Modern Practices

The transition from traditional to modern weed management practices was driven by several factors, including the need for higher agricultural productivity, labor shortages, development of resistance in weeds and the growing complexity of weed infestations. Additionally, advancements in science and technology provided the tools and knowledge necessary to develop more effective solutions for changed weed flora.

Key Drivers

- **Population Growth and Food Demand**: The increasing global population necessitated higher crop production of foodgrains, which traditional methods could not support efficiently.
- **Labor Shortages**: Urbanization and industrialization reduced the availability of agricultural labor, prompting the need for mechanized and automated solutions.
- **Herbicide Resistance**: The emergence of herbicide-resistant weed species required innovative approaches to maintain effective control. Moreover there is continuous shift in weed flora in all field crops due to climatic change.
- **Environmental Concerns**: The negative impacts of some traditional methods, such as soil erosion and water pollution, highlighted the need for sustainable alternatives.
- **Technological Advancements**: Innovations in chemistry, biotechnology, and precision agriculture enabled the development of modern weed control tools and techniques.

Modern Weed Management Practices

Modern weed management practices integrate advanced technologies, scientific principles, and sustainable approaches to achieve effective and efficient weed control. These methods are designed to address the limitations of traditional practices while meeting the demands of contemporary agriculture.

Key Features of Modern Practices

- **Chemical Control**: The use of herbicides has revolutionized weed management. Modern herbicides are designed for specific weed species, minimizing damage to crops and the environment. The selective herbicide should not only be safe to the present crop but the succeeding crop should not be damaged with its residual effects.

- **Biological Control**: This approach involves the use of natural enemies, such as insects, fungi, or bacteria, to suppress weed populations. For instance, the rust fungus *Puccinia chondrillina* has been used to control skeleton weed (*Chondrilla juncea*) in Australia.
- **Mechanical and Robotic Weeding**: Advanced machinery, including automated weeders and robotic systems, has replaced traditional tools. These technologies use sensors and artificial intelligence to identify and remove weeds with precision.
- **Cultural and Preventive Practices**: Modern crop management strategies, such as cover cropping, conservation tillage, and precision planting, are employed to reduce weed pressure.
- **Integrated Weed Management (IWM)**: IWM combines multiple control methods, including chemical, biological, mechanical, and cultural approaches, to achieve sustainable weed management even of problematic weeds

Technological Innovations

- **Precision Agriculture**: Technologies like GPS-guided equipment, drones, and remote sensing allow farmers to monitor and manage weeds with high accuracy. For example, drones equipped with multispectral cameras can detect weed infestations early, enabling targeted interventions.
- **Genetic Engineering**: The development of herbicide-tolerant genetically modified (GM) crops, such as glyphosate-resistant soybeans, has facilitated effective weed control while reducing the cost of cultivation.
- **Data-Driven Decision Making**: Digital platforms and software tools provide farmers with real-time data on weed populations, herbicide efficacy, and environmental conditions, enabling informed decision-making.

Comparison of Traditional and Modern Practices

The Table below summarizes the key differences between traditional and modern weed management practices:

Aspect	Traditional Practices	Modern Practices
Labor Requirement	High	Low
Effectiveness	Limited	High
Environmental Impact	Often negative (*e.g.*, soil erosion)	Reduced (*e.g.*, targeted herbicide application)
Scale of Application	Small-scale	Large-scale
Technological Input	Minimal	High (*e.g.*, robotics, precision tools)

Case Study: Adoption of Modern Practices in the United States

The United States has been at the forefront of adopting modern weed management practices. The widespread use of herbicide-tolerant GM crops, such as Roundup Ready corn and soybeans, has significantly reduced the reliance on traditional tillage methods. Additionally, precision agriculture tools, such as variable-rate sprayers, have enabled farmers to apply herbicides more efficiently, reducing costs and environmental impact (Heap, 2014).

Challenges and Future Directions

While modern weed management practices offer numerous advantages, they are not without challenges. The over-reliance on chemical herbicides has led to the proliferation of herbicide-resistant weeds, necessitating the development of new modes of action and integrated approaches. With continuous use of one group of herbicides, there is a shift in weed flora from traditional to new weed species. Furthermore, the high cost of advanced technologies may limit their adoption by small-scale farmers.

Future research and innovation should focus on developing cost-effective, sustainable, and scalable solutions. For example, advancements in artificial intelligence and machine learning could further enhance the precision and efficiency of robotic weeders.

C. Importance of Adopting Modern Weed Management Methods

Weeds are among the most persistent and detrimental challenges in agriculture, competing with crops for essential resources such as light, water, nutrients, and space. Left unmanaged, weeds can significantly reduce crop yields, compromise quality, and increase production costs. In the context of modern agriculture, where efficiency, sustainability, and profitability are paramount, adopting modern weed management methods is no longer optional but essential.

1. Enhancing Agricultural Productivity

Weeds are notorious for their ability to outcompete crops, often leading to substantial yield losses. Traditional weed control methods, while effective to some extent, are often labor-intensive, time-consuming, and insufficient for large-scale operations. Modern weed management methods, such as precision herbicide application, integrated weed management (IWM), and the use of advanced machinery, offer significant advantages in improving productivity of field crops.

- ✰ **Precision Herbicide Application**: Technologies like GPS-guided sprayers and drone-based herbicide delivery systems enable targeted application of herbicides, reducing wastage and ensuring that weeds are effectively controlled without harming crops.

- **Integrated Weed Management (IWM)**: This approach combines cultural, mechanical, biological, and chemical methods to manage weeds more effectively. For instance, crop rotation and cover cropping can suppress weed growth while maintaining soil health.
- **Yield Preservation**: Studies indicate that modern weed control methods can reduce yield losses by up to 40 per cent in major crops like wheat, maize, and rice (Chauhan *et al.*, 2021).

By adopting these advanced techniques, farmers can ensure optimal resource utilization and maximize their crop yields.

2. Promoting Environmental Sustainability

Traditional weed control practices, particularly excessive reliance on chemical herbicides, have raised concerns about environmental degradation, including soil contamination, water pollution, and loss of biodiversity. Modern weed management methods prioritize sustainability by minimizing the ecological footprint of agricultural practices, such as:

- **Reduced Herbicide Use**: Technologies like weed-detecting sensors and artificial intelligence (AI)-driven weed identification systems allow for site-specific herbicide application, significantly reducing chemical usage.
- **Biological Control**: The use of natural predators, pathogens, or competitors to suppress weed populations is gaining traction as an eco-friendly alternative to chemical herbicides. For example, the introduction of fungal pathogens like *Puccinia chondrillina* to control skeleton weed has shown promising results.
- **Conservation Tillage**: Modern tillage practices, such as no-till or reduced-till farming, help preserve soil structure and organic matter while suppressing weed growth. These practices also reduce greenhouse gas emissions by minimizing soil disturbance.

Adopting these methods aligns with global efforts to promote sustainable agriculture and mitigate climate change impacts.

3. Combating Herbicide Resistance

One of the most pressing challenges in modern agriculture is the rise of herbicide-resistant weed species. Overreliance on a limited number of herbicides has led to the evolution of resistant weed populations, rendering many traditional chemical control methods ineffective. Few alternatives are:

- **Diversified Weed Control Strategies**: Modern weed management emphasizes the integration of multiple control methods to reduce selection pressure on weeds. For instance, alternating herbicides with different modes of action or incorporating non-chemical methods like encouraging crop competition can delay the development of resistance.

- **Herbicide-Resistant Crop Varieties**: Advances in biotechnology have enabled the development of crop varieties resistant to specific herbicides, allowing for effective weed control without harming the crop. However, these technologies must be used judiciously to prevent further resistance issues.
- **Monitoring and Early Detection**: Modern tools like molecular diagnostics and remote sensing facilitate the early detection of herbicide-resistant weed populations, enabling timely intervention.
- **Monitoring Shift in Weed Flora**: With continuous use of one herbicide group, chances of shift in weed flora is very common *i.e.* target weed flora changes to non-target weed flora.

By addressing herbicide resistance proactively, modern weed management methods ensure the long-term efficacy of weed control strategies.

4. Reducing Labor and Operational Costs

Labor shortages and rising operational costs are significant challenges for farmers worldwide. Traditional weed control methods, such as manual weeding, are labor-intensive and often impractical for large-scale farming due to unavailability of required labour at specific time. Modern weed management methods offer cost-effective solutions that reduce the reliance on manual labor.

- **Automation and Robotics**: Autonomous weeders equipped with AI and machine learning can identify and remove weeds with high precision, significantly reducing labor requirements.
- **Economies of Scale**: Technologies like herbicide-resistant crops and precision application systems enable farmers to manage weeds efficiently over large areas, lowering per-unit costs.
- **Long-Term Savings**: While the initial investment in modern weed management technologies may be high, the long-term savings in labor, herbicide use, and yield preservation often outweigh the costs.

These advancements make modern weed management methods particularly appealing for commercial agriculture.

5. Supporting Food Security and Global Demand

The global population is projected to reach 9.7 billion by 2050, necessitating a 70 per cent increase in food production to meet demand (FAO, 2017). Weeds pose a significant threat to food security by reducing crop yields and quality. Modern weed management methods play a crucial role in addressing this challenge.

- **Yield Stability**: By effectively controlling weeds, modern methods ensure consistent and reliable crop production, which is essential for meeting global food demand.
- **Improved Crop Quality**: Weeds can harbor pests and diseases that affect crop quality. For example, parasitic weeds like *Striga* can severely damage cereal crops. The infestation of wheat fields with *Convolvulus arvensis* (Lehli/Hiran Khuri) or *Rumex dentatus* (Kandyali Palak) deteriorates not only quality but also reduces market value of wheat grains Modern methods, such as biotechnological interventions and precision farming, help mitigate these risks.
- **Resource Efficiency**: Advanced weed management techniques optimize the use of inputs like water, fertilizers, and land, ensuring that resources are used efficiently to produce more food with less environmental impact.

By safeguarding crop productivity and quality, modern weed management methods contribute to global food security.

6. Adapting to Climate Change Challenges

Climate change has exacerbated weed problems by altering weed-crop dynamics, expanding the range of invasive species, and increasing the prevalence of herbicide resistance. Modern weed management methods are essential for adapting to these challenges.

- **Climate-Resilient Practices**: Techniques like cover cropping and conservation tillage not only suppress weeds but also enhance soil moisture retention and carbon sequestration, making them suitable for changing climatic conditions.
- **Predictive Modeling**: Advanced tools like climate modeling and weed population forecasting enable farmers to anticipate and prepare for weed outbreaks under different climate scenarios.
- **Invasive Species Management**: Modern methods, such as remote sensing and GIS mapping, help monitor and control the spread of invasive weed species, which are becoming more prevalent due to climate change.

These innovations ensure that weed management strategies remain effective in the face of evolving environmental conditions.

7. Facilitating Precision Agriculture

Modern weed management methods are integral to the broader framework of precision agriculture, which aims to optimize agricultural inputs and outputs through the use of advanced technologies, such as:

- **Data-Driven Decision Making**: Tools like satellite imagery, drones, and IoT sensors provide real-time data on weed distribution and density, enabling farmers to make informed decisions.
- **Variable Rate Technology (VRT)**: This technology allows for the precise application of herbicides, fertilizers, and other inputs based on site-specific conditions, reducing wastage of farm input and improving efficiency.
- **Integration with Digital Platforms**: Modern weed management methods are increasingly integrated with digital farm management platforms, providing farmers with a holistic view of their operations and enabling better resource allocation.

By aligning with precision agriculture principles, modern weed management methods enhance the efficiency and sustainability of farming systems.

D. Core Principles of Modern Weed Management

Effective weed management is a cornerstone of sustainable agriculture, ensuring optimal crop productivity and environmental health. Modern weed management integrates scientific principles, advanced technologies, and ecological approaches to control weed populations while minimizing adverse effects on crops, soil, and ecosystems. The best strategies of modern weed management are

1. Integrated Weed Management (IWM)

Integrated Weed Management (IWM) is a holistic approach that combines multiple weed control methods to achieve sustainable and long-term weed suppression. The principle of IWM is to reduce reliance on any single control method, such as herbicides, and instead use a combination of cultural, mechanical, biological, and chemical practices.

Key Components of IWM

- **Prevention**: Preventing the introduction and spread of weed seeds through measures like cleaning equipment, using certified weed-free seeds, and controlling weed infestations in non-crop areas.
- **Cultural Practices**: Enhancing crop competitiveness through practices such as crop rotation, intercropping, and optimizing planting density.
- **Mechanical Control**: Employing tools like tillage, mowing, and hand weeding to physically remove or suppress weeds.
- **Biological Control**: Utilizing natural enemies of weeds, such as insects, pathogens, or grazing animals, to suppress weed populations.
- **Chemical Control**: Applying herbicides judiciously, with consideration for resistance management and environmental safety.

IWM emphasizes adaptability, requiring continuous monitoring and adjustment of strategies based on weed dynamics, environmental conditions, and cropping systems.

2. Weed Biology and Ecology

Understanding the biology and ecology of weeds is fundamental to modern weed management. This principle involves studying the life cycle, reproductive strategies, and ecological interactions of weed species to design targeted and effective control measures.

Key Aspects

- ☆ **Seed Bank Dynamics**: Weed seed banks in the soil serve as reservoirs for future weed infestations. Strategies like shallow tillage and cover cropping can help deplete seed banks.
- ☆ **Weed-Crop Competition**: Weeds compete with crops for resources such as light, water, space, and nutrients. Early-season weed control is critical, as weeds are most competitive during the crop establishment phase.
- ☆ **Dispersal Mechanisms**: Understanding how weeds spread (*e.g.*, wind, water, animals, or machinery) helps in devising preventive measures.

For example, the biology of Palmer amaranth (*Amaranthus palmeri*), a highly competitive weed, reveals its rapid growth rate and high seed production, necessitating timely and integrated control measures (Ward *et al.*, 2013).

3. Herbicide Stewardship

Herbicides remain a vital tool in modern weed management, but their effectiveness is threatened by the evolution of herbicide-resistant weed populations. Herbicide stewardship involves the responsible use of herbicides to delay resistance development and minimize environmental impacts.

Best Practices for Herbicide Stewardship

- ☆ **Rotating Modes of Action**: Alternating herbicides with different modes of action reduces selection pressure for resistance.
- ☆ **Tank Mixing**: Using herbicide mixtures with multiple effective modes of action can enhance weed control and delay resistance.
- ☆ **Application Timing**: Applying herbicides at the right growth stage of weeds ensures maximum efficacy.
- ☆ **Precision Application**: Technologies like GPS-guided sprayers and variable-rate application systems reduce herbicide waste and off-target effects.

The widespread resistance of weeds like waterhemp (*Amaranthus tuberculatus*) to multiple herbicide classes underscores the importance of stewardship in herbicide use (Heap, 2023).

4. Prevention and Early Intervention

Preventing weed infestations and intervening early are cost-effective and environmentally sustainable principles of modern weed management. These strategies focus on minimizing the establishment and spread of weeds before they become problematic.

Preventive Measures

- ☆ **Sanitation**: Cleaning farm equipment and vehicles to prevent the spread of weed seeds from infested to uninfested fields.
- ☆ **Weed-Free Inputs**: Using certified weed-free seeds, compost, and irrigation water.
- ☆ **Buffer Zones**: Establishing vegetative buffer zones around fields to intercept weed seeds and propagules.

Early Intervention

- ☆ **Scouting and Monitoring**: Regular field scouting to identify weed species and densities at an early stage.
- ☆ **Threshold-Based Control**: Implementing control measures when weed populations exceed economic thresholds.

For instance, early detection and removal of invasive species like giant ragweed (*Ambrosia trifida*) can prevent significant yield losses in crops like soybean.

5. Crop and Soil Health Management

Healthy crops and soils are less susceptible to weed infestations. This principle emphasizes the role of agronomic practices in enhancing crop competitiveness and suppressing weed growth.

Enhancing Crop Competitiveness

- ☆ **Optimal Planting Dates**: Sowing crops at the right (optimum) time to establish a competitive canopy before weeds emerge.
- ☆ **Nutrient Management**: Ensuring adequate nutrient availability for crops (whth drill sowing or spot application) while minimizing nutrient access for weeds.
- ☆ **Cover Cropping**: Using cover crops like rye, cowpea, clover *etc.* to suppress weeds through shading and allelopathy.

Soil Health Practices

- **Reduced Tillage**: Minimizing soil disturbance to reduce the germination of weed seeds from the seed bank.
- **Organic Matter Management**: Increasing soil organic matter to improve soil structure and microbial activity, which can suppress weed growth.

Research shows that cover crops like cereal rye (*Secale cereale*) can reduce weed biomass by up to 90 per cent in no-till systems (Mirsky *et al.*, 2013).

6. Resistance Management

Weed resistance to herbicides and other control methods poses a significant challenge to modern agriculture. Resistance management involves proactive strategies to prevent or delay the evolution of resistant weed populations.

Mechanisms of Resistance

- **Target-Site Resistance**: Mutations in the target site of a herbicide render it ineffective.
- **Non-Target-Site Resistance**: Enhanced metabolism or reduced herbicide uptake in weeds.

Resistance Management Strategies

- **Diverse Control Methods**: Combining chemical, cultural, mechanical, and biological controls to reduce selection pressure.
- **Herbicide Rotation**: Alternating herbicides with different modes of action helps to eliminate or delay resistance.
- **Monitoring and Mapping**: Tracking the spread of resistant weed populations to inform management decisions.

For example, glyphosate-resistant horseweed (*Conyza canadensis*) has become a widespread problem in no-till systems, highlighting the need for integrated resistance management strategies.

7. Precision Weed Management

Advances in technology have enabled the development of precision weed management systems, which optimize weed control by targeting specific areas or populations within a field.

Key Technologies

- **Remote Sensing**: Using drones or satellites to detect weed infestations based on spectral differences between crops and weeds.
- **Robotics and Automation**: Autonomous weeding robots equipped with cameras and sensors for site-specific weed control.

- **Artificial Intelligence (AI)**: Machine learning algorithms to identify weed species and recommend control measures.

Benefits of Precision Weed Management

- Reduced input costs by judicious herbicide use.
- Lower environmental impact through targeted applications.
- Improved efficacy by addressing spatial variability in weed populations.

For instance, AI-powered systems like Blue River Technology's "See and Spray" use computer vision to distinguish weeds from crops and apply herbicides only where needed.

8. Environmental and Economic Sustainability

Modern weed management prioritizes practices that are both environmentally and economically sustainable. This principle ensures that weed control measures do not compromise soil health, water quality, or biodiversity while remaining cost-effective for farmers.

Environmental Considerations

- **Minimizing Off-Target Effects**: Using drift-reducing nozzles and buffer zones to protect non-target plants and water bodies.
- **Promoting Biodiversity**: Encouraging beneficial organisms, such as pollinators and natural enemies of weeds, through habitat management.

Economic Considerations

- **Cost-Benefit Analysis**: Evaluating the economic thresholds of weed control measures to ensure profitability.
- **Long-Term Planning**: Investing in sustainable practices like cover cropping and reduced tillage to reduce future weed management costs.

Studies indicate that integrating cover crops and reduced herbicide use can lower input costs while maintaining or improving crop yields (Liebman *et al.*, 2016).

Chapter 2

Technological Advancements in Weed Management

Technological innovations have revolutionized weed management, offering precision, efficiency, and sustainability. These advancements integrate cutting-edge tools and data-driven approaches to mitigate weed infestations effectively, are:

- ☆ **Precision Agriculture Tools**: GPS-guided systems and drones enable site-specific weed management by mapping weed infestations and applying herbicides only where needed. This reduces chemical usage and minimizes environmental impact. For instance, weed-detection drones equipped with multispectral cameras can identify weed species based on spectral signatures.
- ☆ **Robotics and Automation**: Autonomous weeding robots, such as the "FarmWise Titan," use artificial intelligence (AI) and machine vision to distinguish between crops and weeds, removing the latter mechanically without herbicides.
- ☆ **Herbicide-Resistant Crops**: Advances in genetic engineering have led to crops resistant to specific herbicides, such as glyphosate-resistant soybeans, allowing targeted weed control without harming crops (Duke, 2018).
- ☆ **Digital Weed Identification**: Smartphone apps and AI-based platforms, like "Plantix," assist farmers in identifying weeds and recommending control measures based on real-time data.
- ☆ **Integrated Weed Management (IWM) Software**: Decision-support systems, such as "WeedSOFT," combine weather data, crop growth models, and herbicide efficacy to optimize weed control strategies.

These technologies reduce labor, enhance efficiency, and promote sustainable practices in modern agriculture. Some technological advancements in weed management are

A. Precision Weed Control Technologies

Precision weed control technologies leverage advanced tools and data-driven approaches to target weeds with high accuracy, reducing herbicide use and environmental impact. These technologies include:

- ☆ **Weed Detection Systems**: Cameras and sensors integrated with machine learning algorithms identify weed species in real-time, enabling site-specific treatments (Slaughter *et al.*, 2008).
- ☆ **Autonomous Weed Control Robots**: Robots equipped with GPS and AI selectively apply herbicides or mechanically remove weeds, enhancing efficiency.
- ☆ **Variable Rate Application (VRA)**: Drones and sprayers equipped with GPS and GIS optimize herbicide application rates based on weed density and location (Shaner and Beckie, 2014).

These innovations improve weed management while promoting sustainable agriculture.

B. Robotics and Automation in Weed Management

The integration of robotics and automation in weed management is revolutionizing modern agriculture by enhancing precision, efficiency, and sustainability. These technologies are designed to address labor shortages, reduce herbicide usage, and minimize environmental impact while maintaining high productivity.

Key Applications of Robotics and Automation

1. Autonomous Weed Removal Robots

- ☆ Autonomous robots equipped with advanced sensors, cameras, and artificial intelligence (AI) can identify and remove weeds with remarkable precision. For instance, robots like the "FarmWise Titan" use machine vision to distinguish between crops and weeds, enabling targeted mechanical removal without damaging crops.

2. Precision Spraying Systems

- ☆ Automated sprayers, such as the "Blue River Technology's See and Spray," employ AI and computer vision to detect weeds and apply herbicides only where needed. This reduces chemical input by up to 90 per cent, lowering costs and environmental risks (Liu *et al.*, 2020).

3. Laser-Based Weed Control

- Laser weeding systems, such as those developed by Carbon Robotics (2022), use high-energy lasers to destroy weeds at the cellular level. This chemical-free approach is particularly effective for organic farming systems.

4. Swarm Robotics

- Small, interconnected robots working in swarms can collectively manage large fields. These systems are scalable and adaptable to various terrains, offering a cost-effective solution for weed control.

Advantages of Robotics and Automation

- **Precision:** Advanced algorithms ensure accurate weed identification and removal.
- **Efficiency:** Robots can operate continuously, reducing the need for manual labor.
- **Sustainability:** Minimizes herbicide use and promotes eco-friendly practices.

Challenges and Future Prospects

Despite their potential, high initial costs, limited field adaptability, and the need for skilled operators remain challenges. However, ongoing advancements in AI, robotics, and machine learning are expected to make these technologies more accessible and versatile in the near future.

C. Computer Vision and Artificial Intelligence in Weed Management

The integration of **Computer Vision (CV)** and **Artificial Intelligence (AI)** in modern weed management has revolutionized precision agriculture. These technologies enable real-time weed detection, classification, and targeted control, reducing herbicide usage and improving crop yields.

Applications of Computer Vision in Weed Management

- **Weed Identification and Classification**: CV systems use high-resolution cameras and sensors to capture images of crop fields. Advanced algorithms, such as Convolutional Neural Networks (CNNs), analyze these images to differentiate between crops and weeds based on shape, color, and texture.
- **Weed Density Mapping**: CV systems generate spatial maps of weed distribution, providing actionable insights for site-specific weed control.

- ☆ **Autonomous Weeding Robots**: Robots equipped with CV systems can navigate fields, identify weeds, and perform mechanical or chemical removal with precision.

Role of Artificial Intelligence

AI enhances CV systems by enabling:

1. **Deep Learning Models**: AI-powered models, such as YOLO (You Only Look Once) and Faster R-CNN, achieve high accuracy in weed detection under varying environmental conditions.
2. **Decision Support Systems**: AI algorithms process data from CV systems to recommend optimal weed control strategies, considering factors like weed species, growth stage, and crop type.
3. **Real-Time Adaptation**: AI enables autonomous systems to adapt to dynamic field conditions, such as changes in lighting or weed density.

Recent Advances

- ☆ **Multispectral and Hyperspectral Imaging**: These technologies, combined with AI, improve weed detection accuracy by analyzing spectral signatures invisible to the human eye (Xiang *et al.*, 2021).
- ☆ **Edge Computing**: AI models deployed on edge devices allow real-time weed detection without relying on cloud connectivity, enhancing efficiency in remote areas.

Benefits

- ☆ Reduced herbicide application through site-specific targeting.
- ☆ Minimized crop damage by precise weed control.
- ☆ Enhanced sustainability by lowering environmental impact.

Challenges

- ☆ High initial costs of CV and AI systems.
- ☆ Need for robust training datasets to improve model accuracy.
- ☆ Complexity in distinguishing weeds from crops in mixed-species fields.

D. GPS/GIS and Variable Rate Technology

The integration of GPS (Global Positioning System), GIS (Geographic Information Systems), and Variable Rate Technology (VRT) has revolutionized modern weed management by enabling precision agriculture. These technologies allow farmers to apply herbicides and other inputs with spatial accuracy, reducing costs and minimizing environmental impact.

Role of GPS and GIS in Weed Management

- **GPS Technology**: GPS provides real-time location data, enabling precise mapping of weed infestations. This data helps in creating site-specific weed control strategies, ensuring targeted herbicide application.
- **GIS Mapping**: GIS software processes spatial data collected via GPS to generate detailed weed distribution maps. These maps allow for the identification of weed hotspots and the development of tailored management plans.

Variable Rate Technology (VRT)

VRT enables the application of herbicides at variable rates based on weed density and type. Using GPS and GIS data, sprayers equipped with VRT can adjust the amount of herbicide applied in real-time. This reduces over-application, prevents herbicide resistance, and enhances cost-efficiency.

Applications in Weed Control

- Site-specific herbicide application in fields with varying weed densities.
- Integration with drones for aerial weed mapping and precision spraying.
- Monitoring herbicide effectiveness through post-application GIS analysis.

Scientific Advancements

Recent advancements include the use of machine learning algorithms integrated with GPS/GIS for automated weed detection and classification (Shirzadifar *et al.*, 2022). These innovations further enhance the precision and sustainability of weed management practices.

E. Drones and Remote Sensing

The integration of drones and remote sensing technologies has revolutionized modern weed management by enabling precise, efficient, and cost-effective monitoring and control strategies. These technologies provide real-time, high-resolution data, allowing for early detection and targeted intervention.

Applications in Weed Management

1. Weed Mapping and Identification

Drones equipped with multispectral, hyperspectral, or thermal cameras can capture detailed images of fields. These images are processed using advanced algorithms and machine learning models to differentiate between crops and weeds based on spectral signatures, growth patterns, and canopy structures (Torres-Sánchez *et al.*, 2015).

2. Variable Rate Herbicide Application

By identifying weed-infested zones, drones enable site-specific herbicide application, reducing chemical usage and minimizing environmental impact. This precision agriculture approach enhances sustainability and lowers input costs.

3. Monitoring Herbicide Efficacy

Post-application, drones can assess the effectiveness of herbicide treatments by monitoring weed regrowth and density over time. This data helps in optimizing future weed control strategies.

Advantages of Drone Technology

- ☆ **Cost Efficiency**: Reduces labor and operational costs compared to manual scouting.
- ☆ **Rapid Data Collection**: Covers large areas quickly, providing timely insights.
- ☆ **Environmental Benefits**: Promotes responsible chemical use, reducing runoff and contamination risks.

Challenges and Future Prospects

While drones offer immense potential, challenges such as high initial costs, regulatory restrictions, and the need for skilled operators persist. However, advancements in artificial intelligence and sensor technologies are expected to further enhance their utility in weed management.

F. Sensor-Based Weed Control Systems

Sensor-based weed control systems represent a cutting-edge approach to precision agriculture, leveraging advanced technologies to improve weed management efficiency. These systems utilize sensors to detect, identify, and differentiate weeds from crops in real-time, enabling targeted herbicide application or mechanical removal. By minimizing herbicide use, they reduce environmental impact and operational costs while maintaining crop health and yield.

Key Components of Sensor-Based Systems

1. Sensors

- ☆ **Optical Sensors**: Detect differences in light reflectance between weeds and crops using multispectral or hyperspectral imaging.
- ☆ **LiDAR Sensors**: Measure vegetation height and density to distinguish weeds from crops.
- ☆ **Electromagnetic Sensors**: Identify soil and vegetation properties to locate weeds.

2. Data Processing Units

- Algorithms analyze sensor data to classify plants based on size, shape, and spectral signatures. Machine learning models are increasingly used to improve accuracy.

3. Actuators

- Sprayers or mechanical tools are activated to target weeds precisely, reducing collateral damage to crops.

Applications and Benefits

- **Site-Specific Weed Management**: Sensors enable precise herbicide application only where weeds are detected, reducing chemical usage by up to 90 per cent (Slaughter *et al.*, 2008).
- **Autonomous Weed Control**: Integrated with robotics, sensor-based systems allow fully automated weed removal in fields.

Challenges

- High initial costs and the need for skilled operators limit widespread adoption. However, ongoing advancements in sensor technology and artificial intelligence are expected to address these barriers.

G. Genetic Engineering for Weed Control

Genetic engineering has revolutionized weed management by enabling the development of herbicide-tolerant (HT) crops and bioengineered weed control strategies. HT crops, such as glyphosate-resistant soybean and maize, allow selective herbicide application without damaging crops, significantly improving weed control efficiency. These crops are engineered by introducing genes that confer resistance to specific herbicides, such as EPSPS (5-enolpyruvylshikimate-3-phosphate synthase) genes for glyphosate tolerance. Additionally, RNA interference (RNAi) technology is being explored to target essential weed genes, disrupting their growth and reproduction.

Emerging advancements include gene drives to suppress weed populations by altering reproductive traits (Neve *et al.*, 2018). These innovations promise sustainable weed management solutions.

H. Herbicide-Resistant Crops

Herbicide-resistant crops (HRCs) are genetically engineered or selectively bred plants designed to tolerate specific herbicides, enabling effective weed control without damaging the crop. These crops have revolutionized modern weed management by simplifying herbicide application and reducing labor-intensive practices like mechanical weeding.

Mechanism of Herbicide Resistance (HRC)

HRCs are developed through:

- **Genetic Engineering**: Introduction of genes that confer resistance to specific herbicides. For example, glyphosate-resistant crops (*e.g.*, Roundup Ready® soybeans) contain a gene encoding an altered EPSPS enzyme, which is unaffected by glyphosate [Duke, 2018].
- **Selective Breeding**: Identification and propagation of naturally occurring herbicide-resistant traits in plants.

Advantages of Herbicide-Resistant Crops

- **Enhanced Weed Control**: Broad-spectrum herbicides like glyphosate can be applied post-emergence, effectively managing diverse weed species.
- **Reduced Tillage**: Promotes conservation tillage practices, reducing soil erosion and improving soil health.
- **Increased Yield Potential**: Minimizes weed competition, ensuring optimal crop growth.

Challenges and Concerns

- **Herbicide Resistance in Weeds**: Over-reliance on a single herbicide can lead to the evolution of resistant weed populations, as seen with glyphosate-resistant *Amaranthus* spp. [Heap, 2023].
- **Environmental Impact**: Increased herbicide usage may affect non-target organisms and biodiversity.
- **Regulatory and Public Acceptance**: Concerns over genetically modified organisms (GMOs) persist in some regions.

Examples of Herbicide-Resistant Crops

- **Glyphosate-Resistant Crops**: Soybean, maize, cotton.
- **Glufosinate-Resistant Crops**: Canola, rice.
- **Dicamba-Resistant Crops**: Cotton, soybean.

I. Weed-Suppressive Crops

Weed-suppressive crops are an integral component of integrated weed management (IWM) systems. These crops possess inherent traits or management-induced characteristics that reduce weed growth and competition, thereby minimizing the need for chemical herbicides. The suppression of weeds can occur through various mechanisms, including rapid canopy closure, allelopathy, and competitive resource utilization.

Key Mechanisms of Weed Suppression

1. Rapid Canopy Closure

- Crops like rye (*Secale cereale*) and barley (*Hordeum vulgare*) exhibit fast growth and dense canopy formation, which shades the soil surface and limits light availability for weed germination and growth.
- For example, winter rye can reduce weed biomass by up to 80 per cent when used as a cover crop (Mirsky *et al.*, 2011).

2. Allelopathy

- Certain crops release allelochemicals that inhibit weed seed germination and growth. Sorghum (*Sorghum bicolor*) and black mustard (*Brassica nigra*) are notable examples.
- Studies have shown that sorghum residues can reduce weed emergence by 40–60 per cent due to the release of sorgoleone, a potent allelochemical (Weston and Duke, 2003).

3. Competitive Resource Utilization

- Crops with aggressive root systems, such as alfalfa (*Medicago sativa*), outcompete weeds for water, nutrients, and space.
- High-density planting of crops like maize (*Zea mays*) has also been shown to suppress weed growth by limiting resource availability.

Applications in Cropping Systems

- **Cover Crops**: Weed-suppressive crops like clover (*Trifolium spp.*) and hairy vetch (*Vicia villosa*) are commonly used as cover crops to smother weeds during fallow periods.
- **Intercropping**: Combining weed-suppressive crops with primary crops, such as intercropping maize with cowpea (*Vigna unguiculata*), enhances weed control while maintaining productivity.

Weed-suppressive crops are a sustainable alternative to herbicide reliance, offering environmental and economic benefits. Their integration into cropping systems requires careful selection based on regional weed species, soil conditions, and crop compatibility.

J. Gene Editing Technologies (CRISPR-Cas)

Gene editing technologies, particularly CRISPR-Cas systems, have emerged as transformative tools in modern weed management. CRISPR-Cas (Clustered Regularly Interspaced Short Palindromic Repeats and CRISPR-associated proteins) enables precise genome modification by targeting specific DNA sequences. This technology has been widely adopted in agriculture to develop herbicide-resistant crops and explore innovative weed control strategies.

Mechanism of CRISPR-Cas

- **Targeted DNA Cleavage**: CRISPR-Cas employs a guide RNA (gRNA) to direct the Cas protein (*e.g.*, Cas9) to a specific DNA sequence (Doudna and Charpentier). The Cas protein introduces a double-strand break at the target site.
- **DNA Repair Pathways**: The break is repaired by the plant's natural DNA repair mechanisms, either through non-homologous end joining (NHEJ) or homology-directed repair (HDR). These pathways can result in gene knockouts or precise gene insertions.

Applications in Weed Management

1. **Herbicide Resistance**: Gene editing can create crops with resistance to specific herbicides, allowing selective weed control without harming the crop. For example, CRISPR has been used to modify genes in rice and maize for resistance to glyphosate (Li *et al.*, 2022).
2. **Weed Sterility**: CRISPR can target genes essential for weed reproduction, reducing their ability to propagate.
3. **Weed-Crop Interaction Studies**: CRISPR aids in understanding the genetic basis of weed competitiveness, enabling the development of crops with enhanced competitive traits.

Advantages

- High precision and efficiency.
- Cost-effective compared to traditional breeding.
- Reduced off-target effects with advancements in gRNA design.

Challenges

- Regulatory hurdles in many countries.
- Ethical concerns regarding gene editing in wild weed populations.

Future Prospects

CRISPR-Cas continues to evolve, with newer variants like CRISPR-Cas12 and CRISPR-Cas13 offering expanded applications. These advancements hold promise for sustainable and effective weed management strategies.

K. Nanotechnology in Weed Control

Nanotechnology offers innovative solutions for weed management by enabling precise and efficient herbicide delivery systems. Nano-herbicides, formulated with nanoparticles, enhance the solubility, stability, and bioavailability of active ingredients, reducing environmental contamination and herbicide resistance. These nanoscale carriers, such as liposomes, polymeric nanoparticles, and silica-based nanomaterials, ensure targeted delivery to weed-specific sites, minimizing off-target effects.

Applications of Nanotechnology in Weed Control

- **Targeted Herbicide Delivery:** Nanoparticles encapsulate herbicides, ensuring controlled release and site-specific action on weeds.
- **Smart Weed Sensors:** Nano-sensors detect weed infestations early, enabling timely interventions.
- **Reduced Dosage Requirements:** Enhanced efficacy reduces herbicide quantities, lowering environmental impact.

Recent studies highlight the potential of nano-formulations like mesoporous silica nanoparticles for glyphosate delivery, achieving higher weed control efficiency (Kah *et al.*, 2018). However, further research is needed to address regulatory and safety concerns.

L. Nanoherbicides

Nanoherbicides represent a cutting-edge innovation in weed management, utilizing nanotechnology to enhance the efficacy, precision, and environmental safety of herbicides. These formulations incorporate nanoparticles or nanocarriers to improve the delivery, release, and activity of active ingredients against target weeds. By leveraging nanoscale properties, nanoherbicides address limitations of conventional herbicides, such as poor solubility, drift, and non-target effects.

Key Features of Nanoherbicides

- **Enhanced Bioavailability:** Nanocarriers such as liposomes, polymeric nanoparticles, and silica nanoparticles improve the solubility and absorption of poorly soluble herbicides.
- **Controlled Release:** Nanoformulations allow for sustained and targeted release of active ingredients, reducing the frequency of application and minimizing environmental contamination.
- **Reduced Non-Target Impact:** By improving selectivity, nanoherbicides reduce damage to crops, beneficial organisms, and soil health.
- **Lower Dosage Requirements:** Due to their high efficiency, nanoherbicides require smaller quantities of active ingredients, reducing overall chemical usage.

Examples of Nanoherbicides

- **Atrazine-loaded Nanoparticles:** Studies have shown that atrazine encapsulated in polymeric nanoparticles exhibits higher weed control efficacy and lower leaching potential compared to conventional formulations (De Oliveira *et al.*, 2020).

- **Glyphosate Nanoemulsions:** Nanoemulsions of glyphosate demonstrate improved penetration into weed tissues and reduced volatilization, enhancing weed suppression (Kah *et al.*, 2018a).

Challenges and Future Prospects

While nanoherbicides hold immense potential, challenges such as regulatory approval, cost of production, and potential toxicity to non-target organisms must be addressed. Ongoing research aims to develop biodegradable and eco-friendly nanocarriers to ensure sustainable weed management solutions.

M. Nanosensors for Weed Detection

Advancements in nanotechnology have introduced nanosensors as a revolutionary tool for precise weed detection in modern agriculture. Nanosensors are devices that operate at the nanoscale, capable of detecting minute changes in environmental conditions or specific chemical compounds. Their application in weed management is particularly promising due to their high sensitivity, specificity, and potential for real-time monitoring.

Principle of Operation

Nanosensors function by detecting specific biochemical markers or physical properties unique to weeds. For example, they can identify volatile organic compounds (VOCs) emitted by weed species or detect differences in chlorophyll fluorescence between crops and weeds. These sensors often rely on nanomaterials such as carbon nanotubes, quantum dots, or metallic nanoparticles, which enhance their sensitivity and enable rapid signal transduction.

Applications in Weed Detection

- **Real-Time Monitoring**: Nanosensors can be integrated into autonomous systems like drones or robotic platforms to scan fields and identify weed infestations in real time. This allows for targeted herbicide application, reducing chemical usage and environmental impact.
- **Soil and Air Analysis**: Certain nanosensors are designed to detect weed seeds or VOCs in soil and air, enabling early detection before weeds emerge.
- **Crop-Weed Differentiation**: By analyzing spectral signatures or biochemical markers, nanosensors can distinguish between crops and weeds, even in mixed vegetation.

Advantages

- **Precision**: High sensitivity enables the detection of weeds at early growth stages.
- **Cost-Effectiveness**: Reduces herbicide use and labor costs by enabling site-specific weed management.
- **Environmental Sustainability**: Minimizes chemical runoff and non-target effects on beneficial organisms.

Challenges and Future Directions

Despite their potential, challenges such as high production costs, scalability, and field durability limit widespread adoption. Future research aims to develop cost-effective, robust nanosensors capable of operating under diverse agricultural conditions.

Example

A recent study demonstrated the use of gold nanoparticles functionalized with specific ligands to detect VOCs emitted by *Amaranthus palmeri* (Palmer amaranth), a major weed in row crops. The nanosensor achieved over 90 per cent accuracy in distinguishing this weed from crops in controlled environments (Smith *et al.*, 2022).

Chapter 3

Robotics and Automation in Weed Control

The integration of robotics and automation in weed control represents a transformative shift in modern agriculture, offering precision, efficiency, and sustainability. These technologies leverage advancements in artificial intelligence (AI), computer vision, and sensor systems to identify, target, and manage weeds with minimal human intervention.

Key Applications of Robotics in Weed Control

1. **Autonomous Weed-Removing Robots**: Robots equipped with cameras and AI algorithms can distinguish between crops and weeds. For instance, systems like the "Naïo Dino" robot autonomously navigate fields, mechanically removing weeds without damaging crops.
2. **Precision Spraying Systems**: Automated sprayers use machine vision to detect weeds and apply herbicides only where needed, reducing chemical usage and environmental impact. Examples include the "See and Spray" technology developed by John Deere.
3. **Laser-Based Weed Control**: Some robots employ lasers to destroy weeds at the cellular level. This chemical-free approach is particularly beneficial for organic farming systems.

Advantages of Automation in Weed Management

- ✰ **Reduced Labor Dependency**: Automation addresses labor shortages by performing repetitive tasks efficiently.
- ✰ **Environmental Sustainability**: Targeted weed control minimizes herbicide runoff and promotes biodiversity.

- ☆ **Cost Efficiency**: Long-term savings are achieved through reduced chemical inputs and operational costs.

A. Introduction to Robotics and Automation in Agriculture

The integration of robotics and automation in agriculture has revolutionized traditional farming practices, offering innovative solutions to address labor shortages, improve efficiency, and enhance precision in crop management. In the context of modern weed management, these technologies provide advanced tools to detect, monitor, and control weeds with unparalleled accuracy, reducing reliance on manual labor and chemical herbicides.

Role of Robotics in Agriculture

Robotics in agriculture involves the use of autonomous or semi-autonomous machines designed to perform specific tasks. These machines are equipped with sensors, cameras, and advanced algorithms to navigate fields, identify weeds, and execute targeted actions. Key applications of robotics in weed management include:

- ☆ **Autonomous Weed Removal Robots**: Robots like the "Naïo Dino" and "Ecorobotix" are designed to identify and remove weeds mechanically or through precision spraying. These robots use machine vision and artificial intelligence (AI) to distinguish between crops and weeds, ensuring minimal crop damage.
- ☆ **Precision Spraying Systems**: Robotic sprayers equipped with computer vision can apply herbicides only to weed-infested areas, significantly reducing chemical usage and environmental impact. For example, the "Blue River Technology's See and Spray" system uses AI to detect weeds and spray herbicide with pinpoint accuracy (Shamshiri *et al.*, 2018).
- ☆ **Robotic Cultivators**: These machines use mechanical tools to uproot weeds between crop rows, offering an eco-friendly alternative to chemical control methods.

Automation Technologies in Weed Management

Automation in agriculture encompasses the use of advanced technologies such as AI, machine learning (ML), and the Internet of Things (IoT) to streamline farming operations. In weed management, automation enables:

1. **Weed Detection and Mapping**: Automated systems use drones, satellites, and ground-based sensors to collect high-resolution imagery of fields. These images are analyzed using AI algorithms to identify weed-infested areas and generate weed density maps. This data-driven approach allows for targeted interventions.

2. **Variable Rate Technology (VRT)**: VRT systems integrate weed density maps with automated sprayers to apply herbicides at variable rates, optimizing chemical usage and reducing costs.
3. **Real-Time Monitoring**: IoT-enabled sensors provide real-time data on weed growth, soil conditions, and weather patterns. This information helps farmers make informed decisions about the timing and method of weed control.

Advantages of Robotics and Automation in Weed Management

The adoption of robotics and automation offers several benefits:

- **Labor Efficiency**: Automated systems reduce the dependency on manual labor, addressing labor shortages in agriculture.
- **Environmental Sustainability**: Precision application of herbicides minimizes chemical runoff and soil contamination.
- **Cost-Effectiveness**: Targeted weed control reduces input costs, such as herbicides and fuel.
- **Scalability**: These technologies are scalable and adaptable to various crop types and field sizes.

Challenges and Future Prospects

Despite their potential, robotics and automation face challenges such as high initial costs, the need for technical expertise, and limited adoption in small-scale farming. However, ongoing advancements in AI, sensor technology, and machine learning are expected to make these systems more accessible and affordable in the future. Research efforts are also focused on developing multi-functional robots capable of performing diverse tasks, including planting, harvesting, and weed management, to further enhance agricultural productivity.

B. Advantages of Robotic Weed Control

Robotic weed control has emerged as a transformative solution in modern agriculture, offering numerous advantages over traditional weed management methods. By integrating advanced technologies such as artificial intelligence (AI), machine learning (ML), and precision engineering, robotic systems are revolutionizing how weeds are identified, targeted, and eliminated. Below, we explore the key advantages of robotic weed control, highlighting its potential to enhance agricultural productivity, sustainability, and economic efficiency.

1. Precision Targeting of Weeds

One of the most significant advantages of robotic weed control is its ability to precisely target weeds without affecting surrounding crops. This precision is achieved through advanced sensors, computer vision, and AI algorithms that enable robots to differentiate between weeds and crops in real time.

- **Selective Herbicide Application**: Robotic systems equipped with precision sprayers can apply herbicides only to the weeds, significantly reducing chemical usage. For instance, studies have shown that precision spraying can reduce herbicide application by up to 90 per cent compared to conventional blanket spraying methods (Slaughter *et al.*, 2008).
- **Mechanical Weed Removal**: Some robots use mechanical tools, such as blades or rotary cutters, to physically remove weeds without disturbing the soil or crops. This is particularly beneficial for organic farming systems where chemical herbicides are not permitted.

2. Reduction in Herbicide Usage

Robotic weed control directly addresses the environmental and economic challenges associated with excessive herbicide use. By minimizing the volume of chemicals applied, these systems contribute to sustainable farming practices and reduce the risk of herbicide resistance.

- **Environmental Benefits**: Reduced herbicide usage lowers the risk of chemical runoff into water bodies, mitigating the impact on aquatic ecosystems and biodiversity.
- **Economic Savings**: Farmers benefit from lower input costs, as less herbicide is required. This is especially important given the rising costs of agrochemicals and the increasing demand for eco-friendly farming practices.

3. Labor Efficiency and Cost Savings

The agricultural sector faces a growing labor shortage, particularly in tasks like manual weeding, which are labor-intensive and time-consuming. Robotic weed control offers a viable solution by automating these processes.

- **Reduced Dependency on Manual Labor**: Robots can operate autonomously for extended periods, reducing the need for human intervention. This is particularly advantageous in regions where labor costs are high or where there is a scarcity of skilled workers.
- **Operational Cost Savings**: Although the initial investment in robotic systems may be high, the long-term savings in labor and chemical costs make it a cost-effective solution. For example, autonomous weeding robots like the "FarmWise Titan" have demonstrated the potential to reduce overall weeding costs by up to 50 per cent.

4. Enhanced Crop Yields and Quality

Weed competition is a major factor limiting crop yields, as weeds compete with crops for essential resources such as sunlight, water, and nutrients.

Robotic weed control ensures timely and effective weed management, leading to improved crop performance, due to:

- ☆ **Early Detection and Removal**: Advanced imaging systems enable robots to detect weeds at an early growth stage, preventing them from establishing and competing with crops.
- ☆ **Minimized Crop Damage**: Unlike traditional methods such as broad-spectrum herbicide spraying, robotic systems are designed to avoid harming crops, ensuring higher yields and better-quality produce.

5. Scalability and Adaptability

Robotic weed control systems are highly adaptable and can be scaled to suit various farming operations, from smallholder farms to large commercial enterprises.

- ☆ **Versatility Across Crops**: These systems can be programmed to work with different crop types and field conditions, making them suitable for diverse agricultural landscapes.
- ☆ **Integration with Precision Agriculture**: Robotic weeders can be integrated with other precision agriculture technologies, such as GPS-guided tractors and drones, to create a comprehensive farm management system.

6. Contribution to Sustainable Agriculture

Sustainability is a key focus in modern agriculture, and robotic weed control aligns with this goal by promoting environmentally friendly practices, *i.e.*,

- ☆ **Reduced Soil Disturbance**: Mechanical weeding robots minimize soil disruption compared to traditional tillage methods, preserving soil structure and reducing erosion risks.
- ☆ **Lower Carbon Footprint**: Autonomous robots powered by renewable energy sources, such as solar panels, contribute to reducing the carbon footprint of farming operations.

7. Real-Time Data Collection and Analysis

Robotic weed control systems are equipped with advanced sensors and data collection tools that provide valuable insights into field conditions and weed populations.

- ☆ **Field Mapping**: Robots can generate detailed maps of weed infestations, enabling farmers to implement site-specific management strategies.
- ☆ **Predictive Analytics**: By analyzing historical and real-time data, robotic systems can predict weed emergence patterns, helping farmers plan proactive interventions.

8. Reduced Risk of Herbicide Resistance

The overuse of herbicides has led to the emergence of herbicide-resistant weed species, posing a significant challenge to global agriculture. Robotic weed control addresses this issue by offering non-chemical alternatives.

- ☆ **Mechanical and Thermal Weed Control**: Robots equipped with mechanical tools or thermal technologies, such as lasers or hot water, can eliminate weeds without relying on herbicides, reducing the selection pressure for resistance.
- ☆ **Diverse Weed Management Strategies**: By integrating multiple weed control methods, robotic systems help delay the development of resistance and maintain the efficacy of existing herbicides.

9. Improved Worker Safety

Traditional weed management methods, particularly those involving chemical herbicides, expose farm workers to health risks. Robotic weed control eliminates these risks by automating the process.

- ☆ **Reduced Exposure to Chemicals**: Autonomous robots handle herbicide application, minimizing the need for human contact with potentially hazardous substances.
- ☆ **Safer Working Conditions**: Robots can operate in challenging environments, such as steep terrains or areas with dense weed infestations, reducing the physical strain on workers.

10. Potential for Continuous Operation

Unlike human labor, robotic systems can operate continuously, day and night, under various weather conditions. This capability ensures timely weed management and maximizes operational efficiency, due to:

- ☆ **24/7 Operation**: Equipped with advanced navigation and obstacle detection systems, robots can work autonomously around the clock, increasing the speed and efficiency of weed control.
- ☆ **Weather-Resilient Design**: Modern robotic systems are designed to withstand adverse weather conditions, ensuring consistent performance throughout the growing season.

11. Support for Organic Farming Systems

Robotic weed control is particularly beneficial for organic farming, where the use of synthetic herbicides is prohibited. By providing non-chemical weed management solutions, robots help organic farmers maintain compliance with certification standards.

- ☆ **Mechanical and Thermal Methods**: Robots can employ mechanical weeding tools or thermal technologies to manage weeds effectively in organic systems.
- ☆ **Enhanced Market Access**: By enabling organic weed control, robotic systems help farmers meet the growing consumer demand for organic produce.

12. Technological Advancements and Future Potential

The field of robotic weed control is rapidly evolving, with ongoing advancements in AI, machine learning, and robotics. These innovations are expected to further enhance the efficiency and capabilities of robotic systems.

- ☆ **AI-Driven Improvements**: Machine learning algorithms enable robots to continuously improve their weed identification and targeting accuracy over time.
- ☆ **Integration with IoT**: The Internet of Things (IoT) allows robotic systems to communicate with other farm equipment, creating a connected and intelligent farming ecosystem.

C. Types of Robotics Used in Weed Control

The integration of robotics in agriculture has revolutionized weed management practices, offering precision, efficiency, and sustainability. Robotics in weed control employs advanced technologies such as artificial intelligence (AI), machine vision, and autonomous navigation to identify, target, and eliminate weeds with minimal environmental impact. This section explores the various types of robotics used in weed control, highlighting their features, applications, and advantages.

1. Autonomous Weed-Removing Robots

Autonomous weed-removing robots are self-operating machines designed to identify and remove weeds from agricultural fields without human intervention. These robots rely on advanced sensors, cameras, and machine learning algorithms to distinguish between crops and weeds (Figure 3.1).

Key Features

- ☆ **Machine Vision Systems**: Equipped with high-resolution cameras and image recognition software to identify weeds based on shape, color, and size.
- ☆ **Autonomous Navigation**: Utilize GPS, LiDAR, and inertial measurement units (IMUs) to navigate fields and avoid obstacles.
- ☆ **Mechanical Weed Removal**: Employ tools such as blades, tines, or rollers to uproot or cut weeds at the soil surface.

Figure 3.1. Inter-Row and Intra-Row Robotic Weeder.

Examples

- **EcoRobotix**: A solar-powered robot that uses AI and machine vision to detect and spray herbicides only on weeds, reducing chemical usage by up to 90 per cent.
- **Naïo Technologies' Dino**: An autonomous robot designed for large-scale vegetable farms, capable of mechanical weeding with high precision.

Advantages

- Reduces labor costs by automating weed removal tasks.
- Minimizes herbicide usage, promoting environmental sustainability.
- Operates continuously, even in adverse weather conditions.

2. Spraying Robots

Spraying robots are designed to apply herbicides with precision, targeting only the weeds while sparing the crops. These robots integrate advanced sensors and AI to optimize herbicide application, reducing chemical waste and environmental contamination.

Key Features

- **Selective Spraying Technology**: Detects and sprays herbicides directly onto weeds using nozzles with high accuracy.
- **Variable Rate Application**: Adjusts the amount of herbicide based on weed density and type.

- **Real-Time Data Collection**: Gathers information on weed distribution and field conditions for improved decision-making.

Examples

- **Blue River Technology's See and Spray**: A tractor-mounted robot that uses computer vision and AI to identify weeds and selectively spray herbicides.
- **Ecorobotix AVO**: A lightweight, solar-powered robot capable of ultra-targeted herbicide application.

Advantages

- Reduces herbicide costs by targeting only the weeds.
- Limits environmental impact by avoiding over-application of chemicals.
- Enhances crop health by minimizing herbicide exposure to crops.

3. Laser-Weeding Robots

Laser-weeding robots represent a chemical-free approach to weed control, using high-intensity lasers to destroy weeds at their growing points. These robots are particularly suitable for organic farming systems where herbicide use is restricted.

Key Features

- **Precision Laser Systems**: Emit focused laser beams to thermally damage weed tissues without affecting nearby crops.
- **AI-Based Weed Detection**: Employ deep learning algorithms to differentiate between weeds and crops.
- **Energy Efficiency**: Operate using renewable energy sources such as solar panels.

Examples

- **Carbon Robotics' LaserWeeder**: An autonomous robot that uses AI and lasers to eliminate weeds in row crops with millimeter-level precision. (Figure 3.2)
- **WeedBot**: A laser-weeding system designed for high-value crops, capable of targeting weeds in dense planting arrangements.

Advantages

- Eliminates the need for herbicides, making it ideal for organic farming.
- Reduces soil disturbance compared to mechanical weeding.
- Provides long-term weed control by targeting weed seeds and roots.

Figure 3.2. Carbon Robotics' Laser Weeder.

4. Robotic Cultivators

Robotic cultivators are designed to perform mechanical weeding by tilling the soil around crops to uproot weeds. These robots are equipped with advanced navigation systems and sensors to ensure precise operation.

Key Features

- **Mechanical Weeding Tools**: Use rotating blades, hoes, or tines to disrupt weed growth.
- **Row Guidance Systems**: Navigate between crop rows with high accuracy using GPS and machine vision.
- **Adjustable Depth Control**: Allows customization of tilling depth based on soil type and crop requirements.

Examples

- **FarmDroid FD20**: A solar-powered robotic cultivator that performs both seeding and mechanical weeding in row crops.
- **Naïo Technologies' Oz**: A lightweight robot designed for small-scale farms, capable of weeding between and within rows.

Advantages

- Reduces reliance on herbicides by mechanically removing weeds.
- Improves soil aeration and water infiltration through light tillage.
- Suitable for a wide range of crops, including vegetables, cereals, and legumes.

5. Swarm Robotics

Swarm robotics involves the use of multiple small robots working collaboratively to perform weed control tasks. This approach leverages the collective intelligence of robotic units to enhance efficiency and scalability.

Key Features

- **Distributed Operation**: Multiple robots operate simultaneously, covering large areas in less time.
- **Communication Networks**: Robots communicate and coordinate tasks using wireless technologies such as Bluetooth or Wi-Fi.
- **Scalability**: Additional robots can be deployed to increase coverage and efficiency.

Examples

- **Small Robot Company's Tom, Dick, and Harry**: A trio of robots designed to map, monitor, and manage weeds using AI and swarm intelligence.
- **Swarm Farm Robotics**: A platform that enables multiple robots to work together for tasks such as spraying and mechanical weeding.

Advantages

- Increases operational efficiency by dividing tasks among multiple robots.
- Reduces downtime as individual robots can continue working even if one unit malfunctions.
- Enhances adaptability to different field conditions and crop types.

6. Drone-Based Weed Control

Drones, or unmanned aerial vehicles (UAVs), are increasingly used in weed management for tasks such as weed mapping, herbicide spraying, and monitoring. These aerial platforms offer a bird's-eye view of fields, enabling precise and efficient weed control.

Key Features

- **Aerial Imaging**: Equipped with multispectral or hyperspectral cameras to detect weed infestations based on spectral signatures.
- **Targeted Spraying Systems**: Use precision nozzles to apply herbicides only to affected areas.
- **Real-Time Data Transmission**: Provide instant feedback on weed distribution and field conditions.

Examples

- ☆ **DJI Agras Series**: Agricultural drones designed for precision spraying and weed mapping.
- ☆ **SenseFly eBee X**: A fixed-wing drone used for weed mapping and field analysis.

Advantages

- ☆ Reduces labor and time required for field scouting.
- ☆ Minimizes herbicide usage by targeting specific areas.
- ☆ Enhances decision-making through detailed weed distribution maps.

7. Hybrid Robots

Hybrid robots combine multiple weed control methods, such as mechanical weeding, laser treatment, and selective spraying, into a single platform. These versatile robots are designed to address diverse weed management challenges in a single pass.

Key Features

- ☆ **Multi-Functional Systems**: Integrate mechanical, chemical, and thermal weed control methods.
- ☆ **Customizable Modules**: Allow farmers to adapt the robot's functionality based on field requirements.
- ☆ **Advanced AI Integration**: Use deep learning algorithms to optimize weed control strategies.

Figure 3.3. AgXeed AgBot.

Examples

- **AgXeed AgBot**: A hybrid robot capable of performing mechanical weeding, spraying, and soil preparation (Figure 3.3).
- **RowBot**: A multifunctional robot designed for row crops, offering both mechanical and chemical weed control options.

Advantages

- Provides comprehensive weed control in a single operation.
- Reduces the need for multiple machines, lowering equipment costs.
- Enhances operational efficiency by combining complementary methods.

8. AI-Powered Weed Mapping Robots

AI-powered weed mapping robots are specialized machines designed to survey fields and create detailed weed distribution maps. These robots provide critical data for precision weed management, enabling targeted interventions.

Key Features

- **High-Resolution Imaging**: Capture detailed images of fields for weed identification and mapping.
- **AI-Based Analysis**: Use machine learning algorithms to classify weeds and assess infestation levels.
- **Data Integration**: Sync with farm management software for seamless decision-making.

Examples

- **TerraSentia**: A compact robot designed for field phenotyping and weed mapping in research and commercial applications.
- **AgroBot**: A robot equipped with advanced imaging and AI capabilities for weed detection and mapping.

Advantages

- Improves the accuracy of weed control strategies through detailed mapping.
- Reduces herbicide usage by targeting specific areas.
- Enhances long-term weed management by identifying patterns and trends.

D. Technologies Powering Robotic Weed Control

The integration of robotics into agriculture has revolutionized weed management, offering precision, efficiency, and sustainability. Robotic weed control systems leverage advanced technologies to identify, target, and eliminate weeds while minimizing damage to crops and reducing reliance on chemical herbicides. This section delves into the key technologies that power robotic weed control, providing insights into their functionality and applications.

1. Machine Vision and Image Processing

Machine vision is the backbone of robotic weed control systems, enabling robots to "see" and differentiate between crops and weeds. This technology uses cameras and sensors to capture high-resolution images of the field, which are then analyzed using image processing algorithms.

Key Components

- **Cameras and Sensors**: High-resolution RGB cameras, multispectral cameras, and hyperspectral sensors are commonly used to capture detailed images.
- **Image Processing Algorithms**: Algorithms analyze features such as color, shape, texture, and size to distinguish between crops and weeds.

Applications

- Weed detection and mapping.
- Identification of weed species for targeted control.
- **Example**: A study by Lottes *et al.* (2017) demonstrated the use of convolutional neural networks (CNNs) for real-time weed classification, achieving high accuracy in distinguishing between sugar beet plants and weeds.

2. Artificial Intelligence (AI) and Machine Learning (ML)

AI and ML technologies enable robotic weed control systems to learn and adapt to complex field environments. These systems are trained on large datasets to improve their ability to identify and manage weeds effectively.

Key Features

- **Supervised Learning**: Robots are trained using labeled datasets to recognize specific weed species.
- **Unsupervised Learning**: Systems identify patterns in unlabeled data to detect new or unknown weeds.
- **Reinforcement Learning**: Robots optimize their actions through trial and error to improve weed control strategies.

Applications

- Adaptive weed detection in diverse field conditions.

- Prediction of weed growth patterns for proactive management.
- **Example**: Blue River Technology's "See and Spray" system uses AI to detect and spray herbicides only on weeds, reducing chemical usage by up to 90 per cent (John Deere, 2021).

3. Global Navigation Satellite Systems (GNSS)

GNSS, including GPS, plays a critical role in guiding robotic weed control systems with high precision. These systems rely on satellite signals to navigate fields and execute weed control tasks accurately.

Key Features

- **Real-Time Kinematic (RTK) GPS**: Provides centimeter-level accuracy for precise navigation.
- **Geofencing**: Ensures robots operate within predefined boundaries.

Applications

- Autonomous navigation of robots in large fields.
- Mapping weed infestations for site-specific management.
- **Example**: The Naïo Technologies Oz robot uses GNSS for precise row navigation in vineyards and vegetable fields, ensuring efficient weed removal without damaging crops.

4. Robotics and Mechatronics

The physical design and mechanical systems of robotic weeders are engineered to perform specific tasks such as cutting, uprooting, or spraying weeds. These systems combine robotics and mechatronics for efficient operation.

Key Components

- **Actuators**: Control the movement of robotic arms and tools.
- **End Effectors**: Tools such as blades, grippers, or sprayers used for weed removal.
- **Mobility Systems**: Wheels, tracks, or legs for field navigation.

Applications

- Mechanical weeding using blades or tines.
- Precision spraying of herbicides on targeted weeds.
- **Example**: The Robovator by F. Poulsen Engineering uses mechanical arms equipped with knives to selectively remove weeds in row crops.

5. Weed Mapping and Data Analytics

Weed mapping involves the collection and analysis of spatial data to understand weed distribution patterns. This technology is essential for site-specific weed management and optimizing robotic operations.

Key Features

- **Geospatial Data Collection**: Drones and ground-based sensors collect data on weed infestations.
- **Data Analytics**: Software analyzes weed density, species distribution, and growth stages.

Applications

- Development of weed density maps for targeted control.
- Monitoring the effectiveness of weed management strategies.
- **Example**: The WeedScout system integrates drone imagery with AI-powered analytics to create detailed weed maps, enabling precision management.

6. Internet of Things (IoT) and Connectivity

IoT technology connects robotic weeders to a network of devices, enabling real-time data sharing and remote monitoring. This connectivity enhances the efficiency and scalability of robotic weed control systems.

Key Features

- **Sensors and Actuators**: Collect and transmit data on field conditions and robot performance.
- **Cloud Computing**: Stores and processes data for decision-making.
- **Wireless Communication**: Facilitates remote control and monitoring.

Applications

- Real-time monitoring of robotic operations.
- Integration with farm management systems for coordinated weed control.
- **Example**: The FarmBot system uses IoT technology to remotely control robotic weeders and monitor their performance via a smartphone app.

7. Thermal and Laser Technologies

Thermal and laser technologies offer non-chemical methods for weed control, targeting weeds with precision and minimizing environmental impact.

Key Features

- **Thermal Weeding**: Uses heat from flames or hot water to destroy weed tissues.

- **Laser Weeding**: Employs high-intensity laser beams to target and kill weeds.

Applications

- Organic farming systems where chemical herbicides are restricted.
- Weed control in sensitive environments such as greenhouses.
- **Example**: The Carbon Robotics LaserWeeder uses AI and laser technology to eliminate weeds without disturbing the soil or crops.

8. Energy Systems and Sustainability

Energy efficiency is a critical consideration in the design of robotic weeders. Advances in energy systems ensure that these robots operate sustainably while maintaining high performance.

Key Features

- **Battery Technology**: Lithium-ion batteries provide long-lasting power for field operations.
- **Solar Panels**: Some robots are equipped with solar panels for renewable energy generation.

Applications

- Prolonged operation in remote or off-grid locations.
- Reduction of carbon footprint in agricultural practices.
- **Example**: The Solar Weeder by ecoRobotix operates entirely on solar power, making it an eco-friendly solution for weed control.

9. Integration with Precision Agriculture

Robotic weed control systems are increasingly integrated with precision agriculture technologies, creating a holistic approach to farm management.

Key Features

- **Variable Rate Technology (VRT)**: Adjusts weed control measures based on field variability.
- **Decision Support Systems (DSS)**: Provides recommendations for weed management based on real-time data.

Applications

- Site-specific herbicide application to reduce chemical inputs.
- Optimization of robotic operations for maximum efficiency.
- **Example**: The AgXeed AgBot integrates robotic weed control with precision agriculture tools, enabling seamless data exchange and decision-making.

10. Challenges and Future Directions

While robotic weed control technologies hold immense promise, several challenges must be addressed to ensure widespread adoption.

Challenges

- High initial costs of robotic systems.
- Complexity in training AI models for diverse weed species.
- Dependence on reliable connectivity in remote areas.

Future Directions

- Development of cost-effective robotic solutions for small-scale farmers.
- Advances in AI to improve weed identification accuracy.
- Integration with autonomous farming systems for end-to-end automation.

E. Future Trends and Innovations in Robotics and Automation in Weed Control

The integration of robotics and automation into weed control has emerged as a transformative approach in modern agriculture. These technologies are reshaping how weeds are managed, offering precision, efficiency, and sustainability. This section explores the latest trends and innovations in robotics and automation for weed control, highlighting their potential to revolutionize agricultural practices.

1. Autonomous Weed Control Robots

Autonomous weed control robots are among the most significant advancements in agricultural technology. These robots are equipped with sensors, cameras, and artificial intelligence (AI) algorithms to identify, target, and eliminate weeds with minimal human intervention.

Key Features

- **Computer Vision**: Advanced imaging systems enable robots to distinguish between crops and weeds based on shape, color, and texture.
- **Precision Targeting**: Robots can apply herbicides or mechanical control methods directly to weeds, reducing chemical usage and minimizing environmental impact.
- **GPS Navigation**: High-accuracy GPS systems allow robots to navigate fields autonomously, even in complex terrains.

Examples:

- **Naïo Technologies' Oz and Dino Robots**: These robots are designed for small-scale and large-scale farms, respectively, and use mechanical tools to uproot weeds.
- **Blue River Technology's See and Spray**: This system uses AI-driven cameras to selectively spray herbicides only on weeds, reducing chemical use by up to 90 per cent (Blue River Technology, 2022).

2. AI and Machine Learning in Weed Detection

Artificial intelligence (AI) and machine learning (ML) have become integral to modern weed control systems. These technologies enable robots and automated equipment to make data-driven decisions in real-time.

Applications

- **Weed Identification**: AI algorithms analyze images captured by drones or ground-based robots to identify weed species with high accuracy.
- **Predictive Weed Management**: ML models can predict weed emergence patterns based on environmental data, enabling proactive control measures.

Technological Advancements

- **Deep Learning Models**: Convolutional neural networks (CNNs) are widely used for image recognition tasks, allowing systems to differentiate between crops and weeds with over 95 per cent accuracy.
- **Edge Computing**: On-device processing reduces latency, enabling real-time weed detection and control in the field.

3. Drones for Weed Monitoring and Control

Drones, or unmanned aerial vehicles (UAVs), are becoming indispensable tools for weed management. They offer a bird's-eye view of fields, enabling precise monitoring and targeted interventions.

Capabilities

- **Aerial Mapping**: Multispectral and hyperspectral imaging sensors on drones can detect weed infestations by analyzing vegetation indices like NDVI (Normalized Difference Vegetation Index).
- **Spot Spraying**: Drones equipped with spraying systems can apply herbicides to specific areas, reducing chemical waste and labor costs.

Examples

- **DJI Agras Series**: These drones are designed for agricultural spraying and can cover large areas efficiently.

- **Sentera's FieldAgent**: This platform integrates drone imagery with AI to provide actionable insights for weed management.

4. Automated Mechanical Weed Control

Mechanical weed control has gained renewed interest with the advent of automation. Automated systems use physical tools to remove weeds without relying on herbicides, aligning with organic and sustainable farming practices.

Innovative Technologies

- **Robotic Cultivators**: Machines like the Robovator use cameras and sensors to detect weeds and remove them mechanically while avoiding damage to crops.
- **Laser Weeding**: Emerging systems use high-energy lasers to target and destroy weeds at the cellular level, offering a non-chemical alternative.

Advantages

- Eliminates the risk of herbicide resistance.
- Reduces environmental pollution and chemical residues in crops.

5. Internet of Things (IoT) in Weed Management

The Internet of Things (IoT) is revolutionizing weed management by enabling real-time data collection and connectivity between devices. IoT systems provide farmers with actionable insights to optimize weed control strategies.

Applications

- **Soil and Weather Monitoring**: IoT sensors collect data on soil moisture, temperature, and weather conditions, helping predict weed growth patterns.
- **Smart Sprayers**: IoT-enabled sprayers adjust herbicide application rates based on weed density and environmental conditions.

Example

- **John Deere's Exact Apply Technology**: This system uses IoT sensors to control spray nozzles, ensuring precise herbicide application and reducing waste.

6. Collaborative Swarm Robotics

Swarm robotics, inspired by the collective behavior of insects, involves multiple robots working together to perform weed control tasks. This approach enhances efficiency and scalability in large agricultural fields.

Key Features

- **Decentralized Control**: Each robot operates independently but communicates with others to coordinate actions.
- **Scalability**: Swarm systems can be easily scaled up by adding more robots to the network.

Potential Applications

- Coordinated mechanical weeding across large fields.
- Simultaneous data collection for weed mapping and crop health monitoring.

7. Advances in Sensor Technology

Sensor technology is a cornerstone of robotics and automation in weed control. Recent advancements have improved the accuracy and reliability of weed detection systems.

Types of Sensors

- **LiDAR (Light Detection and Ranging)**: Measures distances using laser light, enabling 3D mapping of fields for precise weed targeting.
- **Hyperspectral Sensors**: Capture a wide range of wavelengths to differentiate between crops and weeds based on their spectral signatures.

Benefits

- Enhanced detection accuracy, even in dense vegetation.
- Ability to identify weeds at early growth stages, improving control efficacy.

8. Integration of Renewable Energy

The use of renewable energy sources, such as solar panels, is becoming increasingly common in robotic weed control systems. This trend aligns with the global push for sustainable agriculture.

Advantages

- Reduces reliance on fossil fuels, lowering carbon emissions.
- Enables continuous operation in remote areas without access to electricity.

Example

- **Solar-Powered Robots**: Companies like EcoRobotix have developed solar-powered weeders that operate autonomously for extended periods.

9. Challenges and Areas for Improvement

While robotics and automation hold immense promise for weed control, several challenges remain:

- ☆ **Cost**: High initial investment limits adoption, particularly for small-scale farmers.
- ☆ **Complexity**: Advanced systems require technical expertise for operation and maintenance.
- ☆ **Regulatory Hurdles**: The use of autonomous equipment and drones is subject to regulatory approvals, which can vary by region.

Addressing these challenges through innovation and policy support will be critical to realizing the full potential of these technologies.

10. Future Outlook

The future of robotics and automation in weed control is poised for rapid growth, driven by advancements in AI, sensor technology, and renewable energy. Emerging trends such as swarm robotics, laser weeding, and IoT integration will further enhance the precision and sustainability of weed management practices. As these technologies become more accessible and affordable, they are expected to play a pivotal role in addressing global challenges such as herbicide resistance, labor shortages, and environmental sustainability.

F. Economic and Environmental Impact of Robotics and Automation in Weed Control

The integration of robotics and automation in weed control has revolutionized modern agriculture. These technologies offer precise, efficient, and sustainable solutions to weed management, addressing both economic and environmental challenges. This section explores the dual impact of robotics and automation on agricultural economics and environmental sustainability, providing insights into their transformative potential.

Economic Impact of Robotics and Automation in Weed Control

1. Cost Efficiency

- ☆ **Reduction in Labor Costs**: Labor shortages and rising wages have made manual weed control increasingly expensive. Automated systems, such as autonomous weeders and robotic sprayers, significantly reduce the need for human labor, lowering operational costs.
- ☆ **Optimized Input Usage**: Precision weed control technologies, such as machine vision and AI-driven sprayers, apply herbicides only where necessary. This reduces herbicide consumption, cutting input costs for farmers.

☆ **Long-Term Investment Benefits**: While the initial investment in robotics can be high, the long-term savings in labor, herbicides, and fuel make these technologies economically viable. For example, autonomous weeders like the Naïo Dino and Blue River Technology's "See and Spray" system have demonstrated significant cost savings over time.

2. **Increased Productivity**

☆ **Enhanced Weed Control Efficiency**: Robotics and automation enable faster and more accurate weed removal compared to manual or traditional methods. This minimizes crop competition and maximizes yields.

☆ **Extended Operational Hours**: Autonomous machines can operate continuously, even at night, increasing the window for weed management and reducing delays caused by labor availability or weather conditions.

3. **Market Accessibility**

☆ **Scalability for Large Farms**: Large-scale farms benefit from the scalability of robotic systems, which can manage vast areas efficiently.

☆ **Potential for Smallholder Adoption**: While initially targeted at large farms, the development of smaller, cost-effective robotic systems is making these technologies accessible to smallholder farmers, fostering inclusivity in agricultural innovation.

Environmental Impact of Robotics and Automation in Weed Control

1. **Reduction in Chemical Usage**

☆ **Targeted Herbicide Application**: Automated systems equipped with AI and machine vision can distinguish between crops and weeds, applying herbicides only where needed. This reduces chemical runoff and minimizes environmental contamination.

☆ **Non-Chemical Weed Control**: Robotics also enables the adoption of non-chemical methods, such as thermal weeding, mechanical weed removal, and laser-based weed destruction, which eliminate the need for herbicides altogether.

2. **Soil Health Preservation**

☆ **Minimized Soil Disturbance**: Traditional mechanical weed control methods, such as tillage, can lead to soil erosion and degradation. Robotic systems, like autonomous weeders, often employ minimal soil disturbance techniques, preserving soil structure and fertility.

- **Reduced Compaction**: Lightweight robotic systems exert less pressure on the soil compared to heavy machinery, reducing soil compaction and promoting better root growth and water infiltration.

3. Lower Carbon Footprint

- **Energy Efficiency**: Many robotic weed control systems are powered by electric or hybrid engines, reducing greenhouse gas emissions compared to conventional fuel-powered machinery.
- **Reduced Herbicide Production Impact**: By lowering herbicide usage, robotics indirectly reduces the environmental footprint associated with herbicide manufacturing, transportation, and application.

4. Biodiversity Conservation

- **Selective Weed Management**: Precision technologies allow for selective removal of invasive weed species while preserving beneficial plants and soil organisms, promoting biodiversity in agricultural ecosystems.
- **Minimized Ecosystem Disruption**: Non-chemical weed control methods reduce the risk of harming non-target organisms, such as pollinators and beneficial insects.

Challenges and Future Prospects

While robotics and automation offer significant economic and environmental benefits, challenges such as high initial costs, technological complexity, and limited access for smallholder farmers remain. However, advancements in AI, machine learning, and sensor technologies are expected to make these systems more affordable and user-friendly in the coming years.

Chapter 4

Computer Vision and Artificial Intelligence in Weed Management

The integration of **Computer Vision (CV)** and **Artificial Intelligence (AI)** has revolutionized modern weed management, offering precision, efficiency, and sustainability. These technologies enable real-time weed detection, species identification, and site-specific herbicide application, reducing chemical usage and environmental impact.

Applications of Computer Vision in Weed Management

- ☆ **Weed Detection and Classification**: CV systems use high-resolution cameras and sensors to capture field images. Advanced algorithms, such as convolutional neural networks (CNNs), process these images to differentiate weeds from crops based on shape, texture, and color.
- ☆ **Real-Time Monitoring**: Drones and autonomous robots equipped with CV systems monitor large fields, identifying weed-infested areas with high accuracy.
- ☆ **Precision Spraying**: CV-guided sprayers apply herbicides only to targeted weeds, minimizing chemical wastage and crop damage.

Role of Artificial Intelligence

- ☆ **Machine Learning Models**: AI-powered models, trained on extensive datasets, improve weed identification accuracy over time. For example, deep learning models can classify weed species with over 90 per cent accuracy (Kamilaris and Prenafeta-Boldú, 2018).
- ☆ **Decision Support Systems**: AI integrates data from CV, soil sensors, and weather forecasts to recommend optimal weed control strategies.

- **Autonomous Weed Control**: Robots like the "See and Spray" system combine AI and CV to autonomously detect and treat weeds in real-time.

Recent Advancements

- **Multispectral and Hyperspectral Imaging**: These technologies enhance weed detection by analyzing light reflectance patterns invisible to the human eye.
- **Edge Computing**: On-device AI processing reduces latency, enabling faster decision-making in the field.

Challenges and Future Directions

Despite their potential, challenges such as high initial costs, data variability, and the need for robust training datasets remain. However, ongoing advancements in AI and CV are expected to address these limitations, making these technologies more accessible to farmers worldwide.

A. Introduction to Computer Vision and AI

The integration of computer vision and artificial intelligence (AI) into agriculture has revolutionized modern weed management practices. These advanced technologies enable precise weed identification, monitoring, and control, reducing the reliance on manual labor and excessive herbicide use. This section explores the fundamentals of computer vision and AI, their applications in weed management, and the scientific principles that underpin their functionality.

What is Computer Vision?

Computer vision is a field of AI that enables machines to interpret and analyze visual data from the environment. By mimicking human vision, computer vision systems extract meaningful information from images or videos and use it for decision-making. In agriculture, this technology is applied to detect and classify weeds, assess crop health, and monitor field conditions.

Key components of computer vision include:

- **Image Acquisition**: Capturing high-resolution images using cameras, drones, or satellites.
- **Image Processing**: Preprocessing techniques such as noise reduction, segmentation, and feature extraction to prepare images for analysis.
- **Object Detection and Classification**: Algorithms that identify and categorize objects (*e.g.*, weeds, crops) within an image.

For example, convolutional neural networks (CNNs), a type of deep learning model, are widely used in computer vision for their ability to recognize patterns and features in complex datasets.

What is Artificial Intelligence (AI)?

AI refers to the simulation of human intelligence in machines, enabling them to perform tasks such as learning, reasoning, and problem-solving. In the context of weed management, AI systems analyze data collected from sensors, cameras, and other sources to make informed decisions. Machine learning (ML), a subset of AI, plays a critical role in training models to recognize weeds and predict their growth patterns.

Types of AI in Weed Management

- **Supervised Learning**: Models are trained using labeled datasets, where images of weeds and crops are annotated. This approach is commonly used for weed identification.
- **Unsupervised Learning**: Algorithms identify patterns in unlabeled data, useful for detecting anomalies in field conditions.
- **Reinforcement Learning**: AI agents learn by interacting with the environment, optimizing actions such as herbicide application.

Applications of Computer Vision and AI in Weed Management

The combination of computer vision and AI has led to significant advancements in precision agriculture. Below are some key applications:

1. Weed Detection and Classification

- AI-powered computer vision systems can differentiate between crops and weeds based on shape, color, and texture. For instance, a CNN model trained on field images can achieve over 90 per cent accuracy in identifying common weed species (Kamilaris and Prenafeta-Boldú, 2018).
- Autonomous robots equipped with cameras and AI algorithms can navigate fields, detect weeds, and remove them mechanically or chemically.

2. Site-Specific Herbicide Application

- Traditional blanket spraying methods often lead to overuse of herbicides, increasing costs and environmental risks. AI-driven systems enable site-specific herbicide application by targeting weeds only where they are present.
- For example, the "See and Spray" technology uses computer vision to detect weeds in real time and applies herbicide precisely, reducing chemical usage by up to 90 per cent (John Deere, 2022).

3. Crop-weed Competition Analysis

- AI models analyze the spatial distribution of weeds and crops to assess competition for resources such as light, water, and nutrients. This information helps farmers make informed decisions about weed control strategies.

4. **Early Weed Detection**

   - ✰ Early-stage weed detection is critical for effective management. AI systems can identify weed seedlings shortly after germination, allowing timely intervention.

Scientific Principles Behind Computer Vision and AI

The effectiveness of computer vision and AI in weed management is rooted in several scientific principles:

- ✰ **Pattern Recognition**: AI models learn to recognize patterns in visual data, such as the unique leaf shapes of different weed species.
- ✰ **Spectral Analysis**: Multispectral and hyperspectral imaging techniques capture data beyond the visible spectrum, enabling the identification of weeds based on their spectral signatures.
- ✰ **Data Fusion**: Combining data from multiple sources (*e.g.*, drones, satellites, ground sensors) enhances the accuracy of weed detection and monitoring.

Challenges and Future Prospects

Despite their potential, the adoption of computer vision and AI in weed management faces challenges such as:

- ✰ **Data Quality**: High-quality, annotated datasets are essential for training AI models.
- ✰ **Scalability**: Implementing these technologies on large-scale farms requires significant investment in hardware and infrastructure.
- ✰ **Adaptability**: AI models must be adaptable to diverse crop and weed species, as well as varying environmental conditions.

Ongoing research aims to address these challenges by developing more robust algorithms, improving data collection methods, and integrating AI with other technologies such as robotics and IoT.

B. Applications of Computer Vision in Weed Control

The integration of computer vision into agricultural practices has revolutionized weed management by enabling precise, efficient, and scalable solutions. Computer vision, a subfield of artificial intelligence (AI), involves the use of algorithms and image processing techniques to analyze visual data. In weed control, this technology is applied to identify, classify, and manage weeds with unprecedented accuracy, reducing the reliance on manual labor and excessive herbicide use. Below, we explore the key applications of computer vision in weed management.

1. Weed Identification and Classification

Computer vision systems are adept at distinguishing between crops and weeds based on their morphological, spectral, and spatial features. These systems utilize advanced imaging techniques, such as multispectral and hyperspectral imaging, to capture data that is not visible to the human eye.

- ☆ **Image Processing Techniques**: Algorithms process images to detect differences in leaf shape, texture, and color. For example, convolutional neural networks (CNNs) are widely used to classify weeds and crops with high accuracy.
- ☆ **Field Applications**: Systems equipped with cameras mounted on drones or tractors can scan fields in real-time, identifying weed species and their spatial distribution. This allows for targeted interventions.

Example: A study by Milioto *et al.* (2018) demonstrated the use of deep learning models to classify weeds in sugar beet fields with an accuracy exceeding 93 per cent. Such advancements highlight the potential of computer vision in automating weed identification tasks.

2. Precision Herbicide Application

One of the most impactful applications of computer vision in weed control is precision herbicide application. By identifying the exact location of weeds, computer vision systems enable site-specific herbicide spraying, reducing chemical usage and environmental impact.

- ☆ **Spot Spraying Systems**: These systems use real-time image analysis to detect weeds and activate nozzles only in areas where weeds are present. This minimizes herbicide wastage and prevents damage to crops.
- ☆ **Integration with Robotics**: Autonomous robots equipped with computer vision and spraying mechanisms can navigate fields, applying herbicides precisely to weed-infested areas.

Example: The Blue River Technology "See and Spray" system, developed by John Deere, uses computer vision to distinguish between crops and weeds, enabling precise herbicide application. This technology has been reported to reduce herbicide use by up to 90 per cent (Blue River Technology, 2021).

3. Weed Density Mapping and Monitoring

Computer vision systems are also employed to create weed density maps, which provide valuable insights into weed infestation levels across a field. These maps enable farmers to make informed decisions about weed management strategies.

- ☆ **Data Collection**: High-resolution images captured by drones or ground-based sensors are processed to determine weed density and distribution patterns.
- ☆ **Decision Support**: Weed density maps can be integrated into farm management software, allowing for the development of customized weed control plans.

Example: Researchers have used computer vision to map weed infestations in wheat fields, helping farmers to optimize their herbicide application rates and reduce costs (Perez-Ortiz *et al.*, 2016).

4. Autonomous Weed Removal

Autonomous weed removal systems represent a significant leap forward in sustainable agriculture. These systems combine computer vision with robotics to physically remove weeds without the need for chemical inputs.

- ☆ **Mechanical Weeding Robots**: Robots equipped with cameras and computer vision algorithms can identify and uproot weeds mechanically. This approach is particularly beneficial for organic farming systems.
- ☆ **Thermal and Laser Weeding**: Some advanced systems use computer vision to guide thermal or laser-based weed removal technologies, offering a non-invasive alternative to herbicides.

Example: The "EcoRobotix" weeding robot uses computer vision to detect weeds and selectively apply micro-doses of herbicides or mechanically remove weeds, significantly reducing labor and chemical inputs.

5. Early Weed Detection

Early detection of weeds is critical for effective management, as it allows farmers to intervene before weeds compete with crops for resources. Computer vision systems enable early-stage weed detection by analyzing seedling morphology and growth patterns.

- ☆ **Seedling Identification**: Algorithms trained on seedling datasets can differentiate between crop and weed seedlings shortly after germination.
- ☆ **Timely Interventions**: Early detection facilitates timely mechanical weeding or herbicide application, improving weed control outcomes.

Example: A study by Lottes *et al.* (2017) demonstrated the use of computer vision for early weed detection in maize fields, enabling precise interventions during the critical early growth stages.

6. Integration with IoT and Big Data

The combination of computer vision with Internet of Things (IoT) devices and big data analytics enhances the scalability and functionality of weed control systems.

- ☆ **IoT Sensors**: IoT-enabled cameras and sensors can continuously monitor fields, providing real-time data for weed detection and management.
- ☆ **Data-Driven Insights**: The integration of computer vision data with big data platforms allows for predictive modeling and trend analysis, helping farmers anticipate weed outbreaks and plan accordingly.

Example: Smart farming platforms like Xarvio integrate computer vision with IoT devices to provide weed management recommendations based on real-time field data.

Challenges and Future Directions

While computer vision has shown immense potential in weed control, certain challenges remain. These include the need for large, annotated datasets to train algorithms, the high cost of advanced imaging equipment, and the complexity of distinguishing weeds in diverse cropping systems. Future research should focus on developing cost-effective, scalable solutions and improving the robustness of computer vision algorithms under varying field conditions.

C. AI Techniques Used in Weed Control

Artificial Intelligence (AI) has emerged as a transformative tool in modern agriculture, enabling precision weed management with greater efficiency, accuracy, and sustainability. By leveraging advanced algorithms, machine learning (ML), computer vision, and robotics, AI-driven weed control systems are helping farmers reduce herbicide use, minimize crop damage, and optimize labor. This section explores the key AI techniques used in weed control, their applications, and the scientific principles underpinning their functionality.

1. Machine Learning for Weed Identification

Machine learning (ML) is one of the most widely used AI techniques in weed control. ML algorithms analyze large datasets to identify patterns and make predictions, which are crucial for distinguishing weeds from crops.(Liakos *et al.*, 2018)

Supervised Learning

Supervised learning algorithms are trained on labeled datasets containing images of crops and weeds. These algorithms learn to classify plants based on features such as leaf shape, color, texture, and size. For example, convolutional neural networks (CNNs), a type of deep learning model, are commonly used for image recognition tasks in agriculture.

- ✰ *Example:* A CNN-based system can differentiate between broadleaf weeds and grass weeds in a wheat field, enabling targeted herbicide application.

Unsupervised Learning

Unsupervised learning algorithms, such as clustering methods, group plants based on similarities without prior labeling. This is useful for detecting unknown weed species or identifying weed hotspots in a field.

Reinforcement Learning

Reinforcement learning (RL) is used in autonomous weed control systems, where an AI agent learns optimal strategies through trial and error. For instance, RL can guide a robotic sprayer to navigate a field and selectively target weeds while avoiding crops.

Scientific Principle: ML models rely on feature extraction and pattern recognition, enabling them to process complex datasets and make accurate predictions. The accuracy of these models improves with the quality and diversity of training data.

2. Computer Vision for Weed Detection

Computer vision, a subfield of AI, enables machines to interpret and analyze visual data from cameras and sensors. In weed control, computer vision systems are used to detect and classify weeds in real-time.

Image Processing

Computer vision systems preprocess images captured by drones, robots, or stationary cameras to enhance features such as edges, contours, and color contrasts. Techniques like thresholding and edge detection are applied to segment weeds from crops.

Object Detection Algorithms

Advanced object detection algorithms, such as YOLO (You Only Look Once) and Faster R-CNN, are used to locate and classify weeds within an image. These algorithms can process high-resolution images quickly, making them suitable for real-time applications.

- ✰ *Example:* A drone equipped with a computer vision system can fly over a soybean field, detect patches of Palmer amaranth, and generate a weed map for precision spraying.

Multispectral and Hyperspectral Imaging

Beyond visible light, computer vision systems can analyze multispectral and hyperspectral images to detect subtle differences in plant reflectance. Weeds and crops often exhibit distinct spectral signatures, which can be used for accurate identification.

Scientific Principle: Computer vision relies on the principles of image analysis and pattern recognition, enabling machines to interpret visual data with high precision. The integration of spectral imaging enhances the ability to distinguish between plant species.

3. Robotics and Autonomous Weed Control Systems

AI-powered robotics is revolutionizing weed management by automating labor-intensive tasks. Autonomous weed control systems combine robotics, computer vision, and machine learning to perform precise weed removal.

Autonomous Weed Sprayers

Robotic sprayers use AI to identify weeds and apply herbicides only where needed. This reduces chemical usage and minimizes environmental impact.

- ☆ *Example:* The "See and Spray" system by John Deere uses computer vision and ML to detect weeds and spray herbicides with pinpoint accuracy.

Mechanical Weed Removal Robots

Some robots are equipped with mechanical tools, such as blades or harrows, to physically remove weeds without chemicals. These systems are particularly useful in organic farming.

Swarm Robotics

Swarm robotics involves multiple small robots working collaboratively to manage weeds across large fields. These robots communicate and share data to optimize their efficiency.

Scientific Principle: Robotics in weed control leverages kinematics, sensor fusion, and AI algorithms to perform precise and autonomous operations. The integration of GPS and LiDAR enhances navigation and spatial awareness.

4. AI-Driven Decision Support Systems

AI is also used to develop decision support systems (DSS) that assist farmers in making informed weed management decisions. These systems analyze data from various sources, such as satellite imagery, weather forecasts, and soil sensors, to provide actionable insights.

Weed Mapping

AI algorithms process satellite or drone imagery to create detailed weed maps, highlighting areas with high weed density. Farmers can use these maps to plan targeted interventions.

Predictive Analytics

Predictive models use historical data and real-time inputs to forecast weed emergence and growth patterns. This helps farmers time their control measures for maximum effectiveness.

- ☆ *Example:* An AI-based DSS can predict the germination of wild oats in a barley field based on soil temperature and moisture data.

Integration with IoT

Internet of Things (IoT) devices, such as soil moisture sensors and weather stations, provide real-time data to AI-driven DSS. This integration enhances the accuracy and relevance of recommendations.

Scientific Principle: DSS relies on data fusion and predictive modeling to generate insights. The use of AI ensures that these systems can process large datasets and adapt to changing conditions.

5. Natural Language Processing for Weed Management

Natural Language Processing (NLP), another AI technique, is used to analyze textual data and provide weed management recommendations.

Knowledge Extraction

NLP algorithms extract relevant information from scientific literature, extension bulletins, and farmer forums to provide evidence-based solutions for weed control.

Chatbots and Virtual Assistants

AI-powered chatbots use NLP to answer farmers' queries about weed identification, herbicide selection, and application techniques. These tools are accessible via smartphones, making them convenient for smallholder farmers.

Scientific Principle: NLP relies on linguistic analysis and machine learning to process and interpret textual data. The use of AI ensures that these systems can handle diverse languages and terminologies.

6. Challenges and Future Directions

While AI techniques offer significant advantages in weed control, several challenges remain:

Data Quality and Availability

The accuracy of AI models depends on high-quality training data. However, obtaining labeled datasets for diverse weed species and cropping systems can be chailenging.

Cost and Accessibility

Advanced AI systems, such as autonomous robots and hyperspectral imaging, may be cost-prohibitive for small-scale farmers.

Regulatory and Ethical Concerns

The use of AI in agriculture raises questions about data privacy, algorithm transparency, and the potential displacement of human labor.

Future research is focused on addressing these challenges through innovations such as open-access datasets, low-cost sensors, and collaborative AI platforms.

D. Real-World Implementations of Computer Vision and AI in Weed Control

The integration of computer vision and artificial intelligence (AI) into modern agriculture has revolutionized weed management practices. These technologies enable precise weed identification, targeted herbicide application, and real-time decision-making, reducing labor costs, environmental impact, and herbicide resistance. Below are some of the most impactful real-world implementations of computer vision and AI in weed control.

1. Automated Weed Identification Systems

Computer vision systems, powered by AI algorithms, are increasingly used to distinguish weeds from crops in real-time. These systems rely on high-resolution cameras and advanced image processing techniques to analyze plant morphology, leaf texture, and spectral properties.

Example: Blue River Technology's "See and Spray" System

Blue River Technology, a subsidiary of John Deere, developed the "See and Spray" system, which uses AI-driven cameras to detect weeds at the leaf level. The system applies herbicides only to the identified weeds, significantly reducing chemical usage by up to 90 per cent (Blue River Technology, 2023).

How It Works

- Cameras mounted on sprayers capture images of plants.
- AI algorithms classify plants as crops or weeds based on pre-trained datasets.
- Targeted nozzles spray herbicide only on weeds, ensuring precision.

This approach not only minimizes herbicide waste but also protects beneficial soil microbes and reduces the risk of herbicide runoff into water bodies.

2. Robotic Weed Control

Robotic weeders equipped with computer vision and AI are transforming weed management by automating labor-intensive tasks. These robots can navigate fields autonomously, identify weeds, and remove them mechanically or chemically.

Example: Naïo Technologies' Autonomous Robots

Naïo Technologies has developed autonomous robots like "Dino" and "Oz" for weed control in vegetable and row crops. These robots use GPS, LiDAR, and computer vision to detect and uproot weeds without damaging crops.

Key Features

- ☆ AI algorithms analyze plant spacing and growth patterns to differentiate weeds from crops.
- ☆ Robots operate continuously, reducing dependency on manual labor.

Such robots are particularly beneficial in organic farming systems, where chemical herbicides are not an option.

3. UAV-Based Weed Mapping and Management

Unmanned Aerial Vehicles (UAVs), commonly known as drones, are increasingly employed for weed detection and mapping. Equipped with multispectral and hyperspectral cameras, drones collect high-resolution images of fields, which are then analyzed using AI models.

Applications

- ☆ **Weed Mapping:** AI processes drone-captured images to create detailed weed distribution maps. These maps guide farmers in applying herbicides only to infested areas, reducing costs and environmental impact.

Example: PrecisionHawk's AI-Powered Drone Solutions

PrecisionHawk uses AI to process aerial images and identify weed infestations. The resulting data is integrated into variable-rate sprayers for site-specific herbicide application.

Advantages

- ☆ Rapid data collection over large areas.
- ☆ Early weed detection, allowing timely intervention.

4. AI-Driven Decision Support Systems

AI-powered decision support systems (DSS) are helping farmers make informed weed management decisions by analyzing field data, weather conditions, and crop growth stages.

Example: Xarvio Digital Farming Solutions

Xarvio's "Field Manager" platform uses AI to provide tailored weed control recommendations. By integrating satellite imagery, soil data, and machine learning models, the system predicts weed pressure and suggests optimal herbicide application rates and timings.

Benefits

- ☆ Reduces over-application of herbicides.
- ☆ Enhances crop yield by ensuring timely weed control.

5. Deep Learning for Weed-Crop Classification

Deep learning, a subset of AI, has shown remarkable success in distinguishing weeds from crops with high accuracy. Convolutional Neural Networks (CNNs) are commonly used for this purpose.

Example: Research on CNN-Based Weed Detection

A study by Milioto *et al.* (2018) demonstrated the use of CNNs to classify weeds and crops in sugar beet fields with over 95 per cent accuracy. The model was trained on thousands of labeled images, enabling precise weed detection even under varying lighting and soil conditions.

Practical Applications

- ☆ Integration into smart sprayers for real-time weed control.
- ☆ Use in robotic weeders for enhanced performance.

6. Challenges and Future Prospects

Despite their potential, the adoption of computer vision and AI in weed control faces challenges such as high initial costs, the need for large annotated datasets, and variability in field conditions. However, ongoing advancements in AI algorithms, sensor technologies, and cloud computing are expected to address these limitations, making these systems more accessible to farmers worldwide.

E. Challenges in Implementing Computer Vision and AI

The integration of computer vision (CV) and artificial intelligence (AI) in modern weed management has shown immense potential in improving precision agriculture. These technologies enable the identification, classification, and targeted control of weeds, reducing the reliance on blanket herbicide applications. However, their implementation is not without challenges. This section explores the key obstacles faced by researchers, agronomists, and agricultural practitioners in adopting CV and AI for weed management.

1. Data Collection and Quality

The success of AI and computer vision systems heavily depends on the availability and quality of data. In weed management, this involves capturing high-resolution images of crops and weeds under varying conditions.

- **Diverse Weed Species and Growth Stages**: Weeds exhibit significant variability in morphology, growth stages, and environmental adaptations. Collecting a comprehensive dataset that represents this diversity is a time-intensive and resource-heavy task.
- **Environmental Variability**: Factors such as lighting, soil type, weather conditions, and crop canopy density can affect image quality and the accuracy of AI models. For instance, shadows or glare in images can lead to misclassification of weeds.
- **Imbalanced Datasets**: In many cases, datasets are skewed toward common weed species, leaving rare or region-specific weeds underrepresented. This imbalance can reduce the generalizability of AI models.

In-text citation: High-quality datasets are critical for training robust AI models, as noted by Kamilaris and Prenafeta-Boldú (2018).

2. Computational Complexity and Resource Requirements

AI models, particularly deep learning algorithms, require significant computational resources for training and deployment.

- **Hardware Limitations**: Many agricultural operations, especially in developing regions, lack access to high-performance computing infrastructure. This limits the ability to train and deploy sophisticated AI models.
- **Energy Consumption**: Training deep neural networks is computationally intensive and consumes substantial energy, raising concerns about sustainability and cost.
- **Real-Time Processing**: For field applications, such as autonomous weed removal systems, AI models must process data in real-time. Achieving this level of performance on portable devices, such as drones or tractors, remains a technical challenge.

3. Model Generalization and Scalability

AI models trained on specific datasets often struggle to generalize across different environments, crops, and weed species.

- **Overfitting**: AI models may perform well on training data but fail to deliver accurate results in real-world scenarios due to overfitting.
- **Regional Variability**: Weed species and cropping systems vary significantly across regions. A model trained in one geographic area may not perform effectively in another without retraining.

- **Scalability Issues**: Scaling AI systems to large farms or diverse cropping systems requires significant customization and adaptation, which can be costly and time-consuming.

4. Integration with Existing Agricultural Practices

Seamlessly integrating AI and CV technologies into traditional farming systems presents logistical and operational challenges.

- **Farmer Training and Adoption**: Many farmers lack the technical expertise to operate AI-based systems. Providing adequate training and support is essential but often overlooked.
- **Compatibility with Equipment**: AI systems must be compatible with existing farm machinery, such as sprayers, drones, or tractors. Retrofitting older equipment can be expensive and technically challenging.
- **Cost of Implementation**: The initial investment required for AI-based weed management systems, including hardware, software, and training, can be prohibitive for small and medium-scale farmers.

5. Regulatory and Ethical Concerns

The use of AI and computer vision in agriculture raises several regulatory and ethical issues.

- **Data Privacy**: Collecting and processing large amounts of field data may raise concerns about data ownership and privacy, particularly when third-party service providers are involved.
- **Algorithm Transparency**: Many AI models, especially deep learning algorithms, operate as "black boxes," making it difficult to explain their decision-making processes. This lack of transparency can hinder trust and adoption.
- **Regulatory Approval**: In some regions, the deployment of AI-driven agricultural systems requires regulatory approval, which can be a lengthy and complex process.

6. Maintenance and Updating of AI Systems

AI and CV systems require regular maintenance and updates to remain effective in dynamic agricultural environments.

- **Evolving Weed Populations**: Weed populations can evolve resistance to herbicides or change their growth patterns over time. AI models must be updated to account for these changes.
- **Software and Hardware Upgrades**: As technology advances, older AI systems may become obsolete, requiring costly upgrades or replacements.

- **Continuous Data Collection**: To improve model accuracy, continuous data collection and retraining are necessary. This ongoing process demands significant time and resources.

7. Limited Access to Expertise

The development and deployment of AI-driven weed management systems require multidisciplinary expertise in agronomy, computer science, and engineering.

- **Shortage of Skilled Professionals**: There is a limited pool of professionals with expertise in both agriculture and AI, making it challenging to design and implement effective systems.
- **Collaboration Barriers**: Effective collaboration between agronomists, data scientists, and engineers is essential but often hindered by differences in technical language and priorities.

8. Economic and Social Barriers

Economic and social factors also play a significant role in the adoption of AI and CV technologies.

- **High Initial Costs**: The cost of AI systems, including hardware, software, and training, can be a significant barrier for smallholder farmers.
- **Unequal Access**: Farmers in remote or underdeveloped regions often lack access to the infrastructure and resources needed to adopt these technologies.
- **Resistance to Change**: Traditional farming communities may be resistant to adopting new technologies due to cultural or practical reasons.

9. Lack of Standardization

The absence of standardized protocols and benchmarks for AI and CV in agriculture complicates their adoption.

- **Inconsistent Metrics**: Different studies and systems use varying metrics to evaluate performance, making it difficult to compare results or establish best practices.
- **Interoperability Issues**: AI systems from different developers may not be compatible, limiting their integration into broader agricultural workflows.

F. Advancements and Future Trends of Computer Vision and AI in Weed Control

The integration of computer vision and artificial intelligence (AI) into agriculture has revolutionized weed control by enabling precise, efficient, and environmentally sustainable practices. These technologies leverage advanced algorithms, machine learning models, and high-resolution imaging to identify, classify, and manage weeds with unprecedented accuracy. This section explores the latest advancements in computer vision and AI for weed control and highlights emerging trends shaping the future of this field.

Advancements in Computer Vision for Weed Control

Computer vision, a subset of AI, focuses on enabling machines to interpret and process visual data. In weed control, it plays a pivotal role in identifying weed species and distinguishing them from crops. Key advancements include:

1. **High-Resolution Imaging and Sensors**
   - The use of high-resolution cameras, multispectral, and hyperspectral sensors has significantly improved weed detection accuracy. These sensors capture detailed data on plant morphology, color, and spectral signatures, enabling precise differentiation between weeds and crops.
   - For example, hyperspectral imaging can detect subtle differences in chlorophyll content and leaf structure, which are often unique to specific weed species.

2 **Deep Learning Models for Weed Identification**
   - Deep learning algorithms, particularly convolutional neural networks (CNNs), have shown remarkable success in classifying weeds. These models are trained on large datasets of annotated images to recognize weed species under varying environmental conditions.
   - For instance, a study by Milioto *et al.* (2018) demonstrated the effectiveness of CNNs in segmenting weeds from crops in real-time, achieving over 90 per cent accuracy in field trials.

3. **Real-Time Weed Detection Systems**
   - Real-time weed detection systems integrate computer vision with robotics or autonomous vehicles. These systems process visual data on-the-go, enabling immediate action such as targeted herbicide application or mechanical weed removal.
   - Autonomous weeders like the "See and Spray" system by Blue River Technology utilize computer vision to selectively spray herbicides only on weeds, reducing chemical usage by up to 90 per cent.

AI-Driven Decision Support Systems

AI enhances weed control by providing actionable insights through predictive analytics and decision support systems. These systems analyze large datasets to optimize weed management strategies.

1. Predictive Weed Mapping

- ✰ AI models can predict weed infestation patterns based on historical data, weather conditions, and soil characteristics. This allows farmers to implement site-specific weed management practices, reducing costs and environmental impact.
- ✰ For example, machine learning algorithms have been used to predict the emergence of Palmer amaranth (*Amaranthus palmeri*), a highly invasive weed, enabling timely interventions.

2. Integration with IoT and Drones

- ✰ The Internet of Things (IoT) and drones equipped with AI-powered cameras are transforming weed monitoring. Drones can survey large fields quickly, capturing high-resolution images that are analyzed by AI to generate weed density maps.
- ✰ These maps guide precision herbicide application, ensuring that only infested areas are treated.

3. Autonomous Weed Control Robots

- ✰ AI-powered robots equipped with computer vision are increasingly being used for mechanical weed removal. These robots identify weeds and remove them using mechanical tools or lasers, eliminating the need for herbicides.
- ✰ Examples include the "EcoRobotix" and "FarmWise" systems, which are designed to operate autonomously in diverse field conditions.

Future Trends in Computer Vision and AI for Weed Control

As technology continues to evolve, several trends are expected to shape the future of weed control:

1. Edge Computing for On-Device Processing

- ✰ Edge computing involves processing data locally on devices rather than relying on cloud servers. This reduces latency and enables real-time weed detection and control, even in remote areas with limited internet connectivity.
- ✰ Future systems may incorporate edge AI chips for faster and more efficient weed identification.

2. Transfer Learning for Improved Model Training

- Transfer learning allows AI models to leverage knowledge from pre-trained datasets, reducing the need for extensive field-specific data collection. This approach can accelerate the deployment of weed detection systems in new regions or for novel weed species.

3. Multi-Modal Data Fusion

- Combining data from multiple sources, such as visual imagery, thermal sensors, and soil moisture sensors, can improve weed detection accuracy. AI algorithms can integrate these data streams to provide a holistic understanding of weed-crop interactions.

4. Sustainable Weed Management Through AI

- Future AI systems will focus on sustainability by optimizing herbicide use, promoting mechanical weed control, and supporting integrated weed management (IWM) practices. These systems will align with global efforts to reduce environmental impact and enhance agricultural resilience.

5. Open-Source Platforms and Collaborative Research

- Open-source AI platforms are fostering collaboration among researchers, enabling the development of more robust and adaptable weed control solutions. For example, platforms like TensorFlow and PyTorch are widely used for training weed detection models.

Challenges and Opportunities

While computer vision and AI offer immense potential, challenges remain in their adoption for weed control:

- **Data Availability and Quality:** High-quality, annotated datasets are essential for training AI models. Efforts to create standardized datasets for diverse crops and weed species are ongoing.
- **Cost and Accessibility:** The high cost of advanced sensors, drones, and AI systems can limit adoption, particularly for smallholder farmers. Future innovations must focus on affordability and scalability.
- **Adaptability to Field Variability:** AI models must account for variations in lighting, weather, and crop growth stages to ensure consistent performance.

Despite these challenges, the rapid pace of technological advancement presents significant opportunities for improving weed management. Continued research and collaboration between agronomists, engineers, and data scientists will drive the development of innovative solutions.

G. Benefits of Computer Vision and AI in Weed Control

The integration of computer vision and artificial intelligence (AI) into modern agriculture has revolutionized weed management practices. These advanced technologies enable precise, efficient, and sustainable weed control, addressing challenges such as herbicide resistance, labor shortages, and environmental concerns. Below, we explore the key benefits of computer vision and AI in weed control, supported by scientific principles and real-world applications.

1. Enhanced Weed Identification and Classification

Computer vision, powered by machine learning algorithms, allows for accurate identification and classification of weed species in real-time. This is achieved through the analysis of high-resolution images captured by drones, robots, or ground-based cameras.

- ✰ **Precision Weed Detection:** AI models, such as convolutional neural networks (CNNs), can differentiate between crops and weeds based on shape, texture, and spectral properties. For instance, multispectral and hyperspectral imaging can detect subtle differences in plant reflectance, enabling the identification of specific weed species even in dense crop canopies.
- ✰ **Adaptability:** AI systems can be trained to recognize region-specific weed species, making them highly adaptable to different agricultural landscapes. For example, a study by Milioto *et al.* (2018) demonstrated the use of deep learning for weed segmentation in sugar beet fields, achieving high accuracy in distinguishing weeds from crops.

2. Targeted Herbicide Application

One of the most significant benefits of computer vision and AI is their ability to enable site-specific weed management (SSWM). This approach minimizes herbicide use by targeting only the areas where weeds are present.

- ✰ **Reduction in Herbicide Usage:** AI-powered sprayers equipped with computer vision systems can identify and spray herbicides only on weeds, leaving the surrounding crops untouched. For instance, the "See and Spray" technology developed by Blue River Technology uses AI to reduce herbicide usage by up to 90 per cent.
- ✰ **Cost Efficiency:** By reducing the volume of herbicides applied, farmers can significantly lower input costs while maintaining effective weed control.
- ✰ **Environmental Benefits:** Targeted application reduces chemical runoff into water bodies and minimizes the impact on non-target organisms, promoting sustainable agricultural practices.

3. Real-Time Monitoring and Decision Support

AI and computer vision facilitate real-time monitoring of weed infestations, providing actionable insights to farmers and agronomists.

- ☆ **Dynamic Mapping:** Drones equipped with computer vision systems can generate detailed weed maps of entire fields. These maps help in understanding weed distribution patterns and planning control measures accordingly.
- ☆ **Decision Support Systems (DSS):** AI-driven DSS can analyze weed data and recommend optimal control strategies, such as the timing of herbicide application or mechanical weeding. For example, the WEEDSCOUT system uses AI to provide weed density estimates and control recommendations.

4. Labor Efficiency and Automation

The adoption of AI and computer vision reduces the dependency on manual labor for weed management, addressing labor shortages and increasing operational efficiency.

- ☆ **Autonomous Weeding Robots:** Robots like the "EcoRobotix" and "Naïo Technologies' Dino" use computer vision to navigate fields and remove weeds autonomously. These robots can operate continuously, reducing the need for human intervention.
- ☆ **Scalability:** Automation allows for the management of large-scale farms with minimal labor input, making it feasible to implement precision weed control across extensive agricultural areas.

5. Early Detection of Herbicide Resistance

AI systems can assist in the early detection of herbicide-resistant weed populations, enabling timely intervention.

- ☆ **Pattern Recognition:** By analyzing weed growth patterns and herbicide response, AI can identify resistance traits before they become widespread. For instance, machine learning models can detect subtle changes in weed morphology or growth rates that indicate resistance.
- ☆ **Proactive Management:** Early detection allows for the implementation of integrated weed management (IWM) strategies, such as crop rotation or mechanical control, to mitigate resistance development.

6. Integration with IoT and Big Data

The synergy between AI, computer vision, and the Internet of Things (IoT) enhances weed control by leveraging big data analytics.

- **Data-Driven Insights:** IoT devices, such as soil sensors and weather stations, provide contextual data that AI systems can use to optimize weed control measures. For example, AI can correlate weed growth with soil moisture levels or temperature patterns to predict infestation risks.
- **Field-Specific Recommendations:** By analyzing historical and real-time data, AI can provide tailored recommendations for each field, improving the effectiveness of weed management practices.

7. Support for Sustainable Agriculture

The use of computer vision and AI aligns with the principles of sustainable agriculture by reducing chemical inputs, conserving resources, and minimizing environmental impact.

- **Soil Health Preservation:** Targeted herbicide application prevents overuse, preserving soil microbial diversity and fertility.
- **Biodiversity Conservation:** By reducing the indiscriminate use of herbicides, these technologies help protect beneficial organisms, such as pollinators and natural weed predators.

8. Scalability Across Cropping Systems

AI and computer vision technologies are versatile and can be applied across various cropping systems, from row crops to orchards.

- **Row Crops:** In crops like maize, soybean, and wheat, AI-powered systems can manage inter-row and intra-row weeds with high precision.
- **Perennial Crops:** In orchards and vineyards, computer vision can identify weeds growing under tree canopies or between vines, enabling targeted control.

9. Continuous Learning and Improvement

AI systems improve over time through continuous learning, making them increasingly effective in weed control.

- **Machine Learning Models:** As more data is collected, machine learning models become better at identifying weeds and predicting their behavior. This iterative improvement ensures that the technology remains relevant and effective.
- **Adaptation to New Challenges:** AI systems can be updated to address emerging challenges, such as new weed species or changing climatic conditions.

10. Economic and Social Benefits

The adoption of AI and computer vision in weed control has broader economic and social implications.

- **Increased Profitability:** By reducing input costs and improving crop yields, these technologies enhance farm profitability.
- **Knowledge Transfer:** The use of AI promotes the adoption of modern agricultural practices, fostering knowledge transfer and skill development among farmers.

Chapter 5

GPS/GIS and Variable Rate Technology

The integration of Global Positioning System (GPS), Geographic Information System (GIS), and Variable Rate Technology (VRT) has revolutionized modern weed management by enabling site-specific and data-driven decision-making. These technologies collectively enhance precision agriculture, optimizing herbicide application while minimizing environmental impact.

GPS and GIS in Weed Management

- **GPS**: GPS provides real-time location data, enabling precise mapping of weed infestations. By geotagging specific areas, farmers can track weed distribution patterns across fields.
- **GIS**: GIS processes spatial data collected via GPS to create detailed weed maps. These maps allow for the identification of weed hotspots and variability in weed density, aiding in targeted interventions.

Variable Rate Technology (VRT)

VRT enables the application of herbicides at variable rates based on weed density and type, as determined by GPS and GIS data. Key benefits include:

- **Optimized Herbicide Use**: Reduces over-application, lowering costs and environmental risks.
- **Improved Efficacy**: Tailors herbicide application to specific weed pressures, enhancing control.
- **Data Integration**: Combines weed maps with sprayer systems for automated, site-specific herbicide delivery.

Applications in Modern Weed Management

- ☆ **Weed Scouting**: GPS-enabled drones and sensors collect real-time weed data for GIS analysis.
- ☆ **Automated Sprayers**: VRT-equipped sprayers adjust herbicide flow rates dynamically, ensuring precise application.

These technologies are pivotal for sustainable weed management, reducing chemical inputs and improving field productivity (Zhang *et al.*, 2022).

A. Introduction to GPS/GIS and Variable Rate Technology (VRT)

The integration of Geographic Information Systems (GIS), Global Positioning Systems (GPS), and Variable Rate Technology (VRT) has revolutionized modern weed management. These technologies enable precision agriculture by providing spatially accurate data, allowing farmers to optimize inputs, reduce costs, and minimize environmental impacts. This section introduces the concepts of GPS, GIS, and VRT, their applications in weed management, and their role in advancing sustainable agriculture.

Global Positioning System (GPS) in Weed Management

GPS is a satellite-based navigation system that provides real-time location data with high accuracy. In agriculture, GPS is used to map fields, track machinery, and collect geospatial data for decision-making. Its application in weed management includes:

- ☆ **Field Mapping**: GPS enables the creation of detailed field maps that identify weed-infested zones. These maps serve as the foundation for site-specific weed control strategies.
- ☆ **Guidance Systems**: GPS-guided machinery ensures precise application of herbicides, reducing overlap and minimizing chemical use.
- ☆ **Data Collection**: GPS devices record georeferenced data on weed distribution, enabling long-term monitoring and analysis.

For example, GPS-equipped drones are increasingly used to survey fields and identify weed hotspots, providing actionable insights for targeted interventions.

Geographic Information Systems (GIS) in Weed Management

GIS is a computer-based tool that integrates, analyzes, and visualizes spatial data. It plays a critical role in weed management by combining multiple data layers, such as soil properties, crop health, and weed distribution, to generate actionable insights. Key applications of GIS include:

- ☆ **Weed Mapping and Monitoring**: GIS software processes georeferenced data to create weed distribution maps. These maps help identify patterns and trends, enabling proactive management strategies.

- **Decision Support Systems**: GIS integrates data from various sources to provide recommendations for site-specific weed control measures.
- **Environmental Impact Assessment**: GIS tools assess the potential environmental impact of herbicide applications, ensuring compliance with regulations and promoting sustainable practices.

For instance, GIS-based models can predict weed emergence based on historical data and environmental conditions, allowing farmers to time their interventions effectively.

Variable Rate Technology (VRT) in Weed Management

VRT is a precision agriculture tool that adjusts the rate of input application (*e.g.*, herbicides, fertilizers) based on spatial variability within a field. By using GPS and GIS data, VRT ensures that inputs are applied only where needed, reducing waste and improving efficiency. Applications of VRT in weed management include:

- **Site-Specific Herbicide Application**: VRT-enabled sprayers apply herbicides at variable rates, targeting weed-infested areas while sparing weed-free zones.
- **Integration with Weed Sensors**: Advanced VRT systems use real-time weed detection sensors to identify and treat weeds selectively.
- **Cost and Resource Optimization**: By tailoring input application, VRT reduces input costs and minimizes environmental risks.

An example of VRT in action is the use of "green-on-brown" sprayers, which detect and spray weeds in fallow fields, significantly reducing herbicide usage (Shaner, 2014).

Synergy Between GPS, GIS, and VRT

The combined use of GPS, GIS, and VRT creates a powerful framework for precision weed management. GPS provides accurate location data, GIS processes and visualizes this data, and VRT translates insights into actionable field operations. Together, these technologies enable farmers to implement site-specific weed control strategies that are efficient, cost-effective, and environmentally sustainable.

B. Applications of GPS in Weed Control

The integration of Global Positioning System (GPS) technology into modern agricultural practices has revolutionized weed management by enabling precision, efficiency, and sustainability. GPS-based weed control offers site-specific solutions, reducing the overuse of herbicides, minimizing environmental impact, and optimizing crop yields. This section explores the diverse applications of GPS in weed control, highlighting its role in precision agriculture and its potential to transform traditional weed management practices.

1. Site-Specific Weed Management (SSWM)

GPS technology is a cornerstone of Site-Specific Weed Management (SSWM), which involves identifying and treating weed infestations at specific locations within a field rather than applying blanket treatments. This targeted approach is made possible through GPS-enabled mapping and data collection.

- **Weed Mapping**: GPS devices, coupled with sensors or drones, are used to map weed infestations across a field. Weed density, species distribution, and growth stages are recorded with high spatial accuracy.

  Example: A farmer can use GPS-enabled drones equipped with multispectral cameras to identify patches of herbicide-resistant weeds in a soybean field.

- **Variable Rate Application (VRA)**: GPS data is integrated with sprayer systems to apply herbicides only where needed and at the appropriate dosage. This reduces chemical use and prevents herbicide resistance.

  Scientific Principle: VRA relies on georeferenced weed maps to control the flow rate of herbicides, ensuring precise application.

In-text citation: Studies have shown that SSWM using GPS reduces herbicide use by up to 40 per cent, enhancing both economic and environmental sustainability (Christensen *et al.*, 2021).

2. Autonomous Weed Control Systems

GPS technology is a critical component of autonomous weed control systems, including robotic weeders and self-driving tractors. These systems rely on GPS for navigation and weed targeting.

- **Robotic Weeders**: Autonomous robots equipped with GPS can navigate fields with centimeter-level accuracy, identifying and removing weeds mechanically or chemically.

  Example: The "FarmDroid FD20" uses GPS to plant crops and later identify and remove weeds between crop rows without human intervention.

- **Self-Driving Tractors**: GPS-guided tractors are used for precision spraying, ensuring that herbicides are applied uniformly and only on targeted areas.

  Benefit: This reduces operator fatigue and increases operational efficiency, especially in large-scale farming.

In-text citation: The adoption of GPS-guided autonomous systems has been linked to a 25 per cent improvement in weed control efficiency compared to manual methods (Smith *et al.*, 2020).

3. Integration with Remote Sensing Technologies

GPS works synergistically with remote sensing technologies, such as drones and satellite imagery, to enhance weed detection and management.

- **Drone-Based Weed Detection**: Drones equipped with GPS and high-resolution cameras capture images of fields, which are analyzed to identify weed infestations.

  Example: Multispectral and hyperspectral imaging can differentiate between crop plants and weeds based on their spectral signatures.

- **Satellite Imagery**: GPS coordinates are used to georeference satellite images, enabling large-scale weed monitoring and management.

  Application: Farmers can monitor weed growth over time and assess the effectiveness of herbicide applications.

In-text citation: Remote sensing combined with GPS has been shown to improve weed detection accuracy by up to 85 per cent in diverse cropping systems (Zhang *et al.*, 2022).

4. Real-Time Kinematic (RTK) GPS for Precision Guidance

Real-Time Kinematic (RTK) GPS enhances the accuracy of weed control operations by providing centimeter-level positioning. This is particularly useful for inter-row cultivation and mechanical weeding.

- **Inter-Row Cultivation**: RTK GPS guides machinery to precisely target weeds between crop rows without damaging the crops.

  Example: In maize fields, RTK GPS ensures that cultivators operate within a 2 cm margin of error, improving weed removal efficiency.

- **Mechanical Weeding**: RTK GPS enables the use of automated mechanical weeders that can operate close to crops, reducing the need for herbicides.

5. Data-Driven Decision Making

GPS technology facilitates the collection and analysis of geospatial data, empowering farmers to make informed decisions about weed management.

- **Historical Weed Data**: GPS allows for the creation of weed infestation maps over multiple growing seasons, helping farmers identify recurring problem areas.

  Application: Farmers can adjust crop rotation or implement targeted weed control strategies in high-risk zones.

- **Integration with Farm Management Systems**: GPS data is integrated into farm management software, enabling seamless planning, execution, and monitoring of weed control activities.

6. Benefits of GPS in Weed Control

The adoption of GPS technology in weed management offers several advantages:

- ☆ **Cost Savings**: Precise application of herbicides reduces input costs.
- ☆ **Environmental Sustainability**: Targeted treatments minimize chemical runoff and soil contamination.
- ☆ **Labor Efficiency**: Automated systems reduce the need for manual labor.
- ☆ **Improved Crop Yields**: Effective weed control ensures optimal resource availability for crops.

7. Challenges and Future Prospects

While GPS technology has transformed weed management, certain challenges remain:

- ☆ **High Initial Costs**: The adoption of GPS-enabled equipment can be expensive for small-scale farmers.
- ☆ **Skill Requirements**: Farmers need training to effectively use GPS-based systems.
- ☆ **Integration with Emerging Technologies**: The future of GPS in weed control lies in its integration with artificial intelligence (AI) and machine learning (ML) for predictive weed management.

Emerging innovations, such as AI-driven weed identification and real-time herbicide optimization, are expected to further enhance the capabilities of GPS in weed control.

C. Role of GIS in Weed Control

Geographic Information Systems (GIS) have emerged as a transformative tool in modern agriculture, offering precision and efficiency in weed management. By integrating spatial data with advanced analytical tools, GIS enables farmers, researchers, and agronomists to monitor, map, and manage weed infestations effectively. This section explores the role of GIS in weed control, highlighting its applications, advantages, and integration with other technologies.

1. Overview of GIS in Agriculture

GIS is a computer-based system designed to capture, store, analyze, and visualize spatial and geographic data. In agriculture, GIS is widely used for precision farming, including soil analysis, crop monitoring, and pest and weed management. By providing spatial insights, GIS allows for site-specific interventions, reducing resource wastage and environmental impact.

2. Applications of GIS in Weed Control

GIS plays a pivotal role in various aspects of weed management. Below are some of its key applications:

2.1 Weed Mapping and Monitoring

- GIS enables the creation of detailed weed distribution maps by integrating data collected from field surveys, drones, and remote sensing technologies.
- These maps provide insights into the spatial distribution and density of weed populations, helping farmers prioritize areas for intervention.
- For example, researchers have used GIS to map invasive species like *Parthenium hysterophorus* and *Amaranthus palmeri*, identifying hotspots for targeted control measures (Sharma *et al.*, 2021).

2.2 Predictive Weed Modeling

- GIS, combined with predictive modeling, helps forecast weed growth and spread based on environmental variables such as soil type, temperature, and rainfall.
- By analyzing historical and real-time data, GIS can predict the emergence of weeds, enabling preemptive action.
- For instance, GIS-based models have been used to predict the spread of herbicide-resistant weeds in the United States, aiding in the development of resistance management strategies.

2.3 Site-Specific Weed Management

- GIS facilitates site-specific weed control by integrating spatial data with precision agriculture tools like variable-rate sprayers.
- Weed-infested zones can be identified and treated with precise herbicide application, minimizing chemical usage and reducing costs.
- This approach is particularly effective in large-scale farming operations, where uniform treatment is neither practical nor economical.

2.4 Integration with Remote Sensing

- GIS works synergistically with remote sensing technologies, such as satellite imagery and UAVs (Unmanned Aerial Vehicles), to detect and monitor weeds.
- High-resolution imagery can differentiate between crops and weeds based on spectral signatures, which are then analyzed in GIS platforms.
- For example, hyperspectral imaging combined with GIS has been used to detect *Avena fatua* (wild oats) in wheat fields with high accuracy.

3. Advantages of GIS in Weed Control

The adoption of GIS in weed management offers several benefits:

- **Precision and Accuracy**: GIS provides detailed spatial data, enabling precise interventions and reducing the risk of over- or under-treatment.
- **Cost-Effectiveness**: By targeting specific areas, GIS reduces the amount of herbicide required, lowering input costs.
- **Environmental Sustainability**: Site-specific applications minimize chemical runoff and environmental contamination.
- **Data Integration**: GIS can integrate multiple data sources, such as soil maps, weather data, and crop growth models, for holistic weed management.
- **Real-Time Decision Making**: With advancements in real-time data collection, GIS allows for timely interventions, reducing the impact of weed infestations.

4. Challenges and Limitations

While GIS offers numerous advantages, its implementation in weed control is not without challenges:

- **High Initial Costs**: The setup of GIS infrastructure, including software, hardware, and skilled personnel, can be expensive.
- **Data Accuracy**: The effectiveness of GIS depends on the accuracy of input data, which may vary due to inconsistencies in field surveys or remote sensing outputs.
- **Technical Expertise**: Effective use of GIS requires specialized knowledge in spatial analysis and data interpretation, which may not be readily available in all farming communities.
- **Integration with Other Systems**: Seamless integration of GIS with other precision agriculture tools and technologies remains a technical challenge.

5. Future Prospects

The role of GIS in weed control is expected to expand with advancements in technology. Emerging trends include:

- **Machine Learning Integration**: Combining GIS with machine learning algorithms can enhance weed identification and predictive modeling.
- **IoT and Real-Time Data**: The Internet of Things (IoT) can provide real-time data on weed growth, which can be analyzed in GIS platforms for immediate action.
- **Cloud-Based GIS**: Cloud computing allows for the storage and analysis of large datasets, making GIS more accessible to small-scale farmers.

6. Case Study: GIS in Herbicide Resistance Management

A notable application of GIS is in managing herbicide-resistant weeds. In Australia, GIS was used to map the spread of glyphosate-resistant *Lolium rigidum* (rigid ryegrass). By analyzing spatial data, researchers identified resistance hotspots and recommended alternative herbicide strategies, reducing the risk of further resistance development (Powles and Yu, 2010).

D. Variable Rate Technology (VRT) in Weed Management

Variable Rate Technology (VRT) is a precision agriculture tool that enables site-specific weed management by applying inputs such as herbicides at variable rates across a field. This approach contrasts with traditional uniform application methods, offering economic, environmental, and agronomic benefits. By leveraging advancements in sensors, GPS, and data analytics, VRT has become a cornerstone of modern weed management strategies.

Principles of Variable Rate Technology in Weed Management

VRT operates on the principle of spatial variability, recognizing that weed density, species composition, and soil characteristics vary within a field. By tailoring herbicide applications to these variations, VRT minimizes over-application and reduces input costs while maintaining effective weed control.

Key components of VRT include:

- ☆ **Data Collection:** High-resolution data on weed distribution is collected using remote sensing, drones, or ground-based sensors.
- ☆ **Prescription Maps:** Data is processed to create prescription maps that guide the application of herbicides at specific rates for different zones.
- ☆ **Application Technology:** Advanced machinery equipped with GPS and variable-rate controllers adjusts herbicide application rates in real time.

Technologies Enabling VRT in Weed Management

1. Sensors for Weed Detection

Modern VRT systems rely on sensors to detect weed presence and density. These sensors can be classified into two main categories:

- ☆ **Optical Sensors:** These use reflectance properties of plants to differentiate between crops and weeds. For example, the WeedSeeker® system employs near-infrared (NIR) technology to identify green vegetation against bare soil.
- ☆ **Multispectral and Hyperspectral Sensors:** Mounted on drones or satellites, these sensors capture data across multiple wavelengths to identify weed species based on their spectral signatures.

2. GPS and GIS Integration

Global Positioning System (GPS) and Geographic Information System (GIS) technologies are integral to VRT. GPS ensures precise geolocation of weed patches, while GIS processes spatial data to generate weed density maps. These maps are then used to create herbicide application prescriptions.

3. Variable-Rate Applicators

Modern sprayers equipped with variable-rate controllers can adjust the flow rate of herbicides in real time. These systems use solenoid valves or pulse-width modulation (PWM) to vary the application rate based on the prescription map. For instance, a high-density weed patch may receive a higher dose, while areas with low weed pressure receive minimal or no application.

Applications of VRT in Weed Management

1. Site-Specific Herbicide Application

VRT enables precise herbicide application, reducing chemical use and environmental impact. For example, in a study on cornfields, VRT reduced herbicide usage by up to 30 per cent without compromising weed control efficacy (Shaner, 2014a).

2. Management of Herbicide Resistance

By targeting specific weed patches and varying herbicide modes of action, VRT helps delay the development of herbicide resistance. For instance, applying a higher dose to resistant weed populations while reducing application in susceptible areas can slow resistance evolution.

3. Integration with Crop Scouting

VRT can be integrated with digital crop scouting platforms to provide real-time updates on weed dynamics. This allows for adaptive management, where herbicide prescriptions are updated based on current field conditions.

Advantages of VRT in Weed Management

Economic Benefits

- ✰ Reduces herbicide costs by applying chemicals only where needed.
- ✰ Increases crop yield by minimizing crop injury from over-application.

Environmental Benefits

- ✰ Decreases chemical runoff into water bodies, reducing environmental pollution.
- ✰ Promotes sustainable farming practices by optimizing input use.

Agronomic Benefits

- Enhances weed control efficacy by targeting specific weed species and densities.
- Reduces the risk of herbicide resistance through diversified application strategies.

Challenges and Limitations

Despite its advantages, VRT faces several challenges:

- **High Initial Investment:** The cost of VRT equipment, including sensors, GPS, and variable-rate applicators, can be prohibitive for small-scale farmers.
- **Data Accuracy:** The effectiveness of VRT depends on the accuracy of weed detection and mapping. Errors in data collection or processing can lead to suboptimal herbicide application.
- **Complexity:** Implementing VRT requires technical expertise in data analysis, equipment calibration, and prescription map generation.
- **Compatibility Issues:** Ensuring compatibility between different VRT components, such as sensors and sprayers, can be challenging.

Case Studies and Real-World Applications

1. VRT in Soybean Fields

A study conducted in the United States demonstrated the effectiveness of VRT in soybean fields. By using drone-mounted multispectral sensors to map weed density, researchers achieved a 25 per cent reduction in herbicide use compared to uniform application methods (Mulla, 2013).

2. Adoption in Europe

In Europe, VRT has been widely adopted for managing blackgrass (Alopecurus myosuroides) in wheat fields. Farmers use prescription maps generated from satellite imagery to apply herbicides selectively, reducing costs and environmental impact.

Future Prospects

The future of VRT in weed management lies in the integration of artificial intelligence (AI) and machine learning (ML). These technologies can enhance weed detection accuracy and enable predictive modeling for proactive weed management. Additionally, advancements in autonomous machinery, such as robotic sprayers, are expected to further revolutionize VRT applications.

E. Technological Integration for GPS/GIS and Variable Rate Technology in Weed Control

The integration of Geographic Positioning Systems (GPS), Geographic Information Systems (GIS), and Variable Rate Technology (VRT) has revolutionized modern weed management practices. These technologies enable precision agriculture, allowing farmers to optimize inputs, reduce environmental impacts, and improve weed control efficiency. This section explores the principles, applications, and benefits of these technologies in weed management.

1. Overview of GPS and GIS in Weed Management

GPS Technology in Agriculture

GPS technology provides accurate geospatial positioning, enabling farmers to map fields, monitor crop health, and implement site-specific weed control strategies. By using satellite signals, GPS devices can pinpoint locations within a few centimeters, facilitating precise application of herbicides and other inputs.

Key applications of GPS in weed control include:

- ☆ **Mapping Weed Infestations:** GPS-enabled devices can record the exact location of weed patches, creating spatial maps for targeted interventions.
- ☆ **Guidance Systems:** GPS-based auto-steering systems ensure accurate navigation of machinery, reducing overlap and minimizing herbicide wastage.
- ☆ **Data Logging:** GPS allows real-time recording of field operations, enabling detailed analysis of weed control measures.

GIS Technology in Weed Management

GIS is a powerful tool for analyzing spatial data and visualizing weed distribution patterns. It integrates data from multiple sources, such as satellite imagery, drone surveys, and field sensors, to create detailed weed maps.

Applications of GIS in weed control include:

- ☆ **Weed Density Mapping:** GIS software can process data to generate maps showing weed density and distribution across fields.
- ☆ **Predictive Modeling:** By analyzing historical data, GIS can predict weed emergence patterns, helping farmers plan preemptive control measures.
- ☆ **Decision Support Systems:** GIS-based platforms provide actionable insights for selecting appropriate herbicides, application rates, and timing.

2. Variable Rate Technology (VRT) in Weed Control

Principles of VRT

Variable Rate Technology (VRT) enables the site-specific application of inputs, such as herbicides, based on real-time field conditions. VRT systems use sensors, GPS data, and pre-programmed algorithms to adjust application rates dynamically.

Key components of VRT include:

- **Sensors:** Detect weed presence, density, and species in real-time.
- **Control Systems:** Regulate the flow of herbicides to match the detected weed pressure.
- **Actuators:** Mechanically adjust application equipment, such as nozzles or sprayers.

Types of VRT in Weed Management

- **Map-Based VRT:** Uses pre-generated weed maps from GIS to guide herbicide application. This approach is effective for fields with well-documented weed histories.
- **Sensor-Based VRT:** Relies on real-time data from optical or spectral sensors to detect weeds and adjust application rates instantly. For example, WeedSeeker® and Weed-It® systems use near-infrared sensors to identify green vegetation against bare soil.

3 Integration of GPS, GIS, and VRT for Precision Weed Control

The synergy of GPS, GIS, and VRT technologies forms the backbone of precision weed management. This integration allows for the seamless collection, analysis, and application of data to optimize weed control strategies.

Workflow of Integrated Weed Control

- **Data Collection:** GPS and sensors collect geospatial and real-time weed data.
- **Data Analysis:** GIS processes the data to generate weed maps and actionable insights.
- **Application:** VRT systems use the processed data to apply herbicides precisely where needed.

Benefits of Integration

- **Reduced Input Costs:** Targeted herbicide application minimizes wastage, reducing costs.
- **Environmental Sustainability:** Precision application reduces chemical runoff and soil contamination.

- **Improved Efficacy:** Site-specific treatments ensure better control of weed infestations.
- **Time Efficiency:** Automated systems reduce the time required for field operations.

Case Study Example

A study conducted in the United States demonstrated that integrating GPS, GIS, and VRT reduced herbicide use by 30 per cent while maintaining effective weed control (Shaner and Beckie, 2021). This approach also improved crop yields by minimizing competition from weeds.

4. Challenges and Future Prospects

Challenges

Despite its advantages, the adoption of GPS/GIS and VRT in weed management faces several following challenges:

- **High Initial Costs:** The equipment and software required for these technologies can be expensive for small-scale farmers.
- **Technical Expertise:** Farmers need training to operate and interpret data from these systems.
- **Data Accuracy:** The effectiveness of these technologies depends on the quality of input data, which can be affected by sensor calibration and environmental factors.

Future Prospects

Advancements in artificial intelligence (AI) and machine learning (ML) are expected to enhance the capabilities of GPS, GIS, and VRT systems. For instance:

- **AI-Driven Weed Identification:** AI algorithms can improve the accuracy of weed detection and classification.
- **Autonomous Machinery:** Self-driving tractors equipped with GPS and VRT systems can perform weed control operations with minimal human intervention.
- **Integration with IoT:** Internet of Things (IoT) devices can provide real-time data on weed growth, soil conditions, and weather, further enhancing decision-making.

5. Practical Considerations for Implementation

Farmers and agronomists aiming to adopt GPS, GIS, and VRT for weed control should consider the following:

- **Equipment Selection:** Choose systems compatible with existing machinery and farm size.

- **Data Management:** Invest in software capable of processing and analyzing large datasets.
- **Training:** Provide operators with adequate training to maximize the benefits of these technologies.
- **Cost-Benefit Analysis:** Evaluate the economic feasibility of adopting these technologies based on farm-specific conditions.

Benefits of GPS/GIS and VRT in Weed Control

The integration of advanced technologies such as Global Positioning Systems (GPS), Geographic Information Systems (GIS), and Variable Rate Technology (VRT) has revolutionized modern weed management. These tools enable precision agriculture, allowing farmers and agronomists to optimize weed control strategies while minimizing environmental impact and reducing costs. Below are the key benefits of these technologies in weed control.

1. Enhanced Weed Mapping and Monitoring

GPS and GIS technologies enable precise mapping and monitoring of weed infestations, providing a spatial understanding of weed distribution across fields. This is critical for implementing site-specific weed management strategies.

- **Accurate Weed Distribution Data**: GPS-equipped machinery can record the exact location of weed patches during field scouting or automated weed detection using sensors. GIS software then processes this data into detailed weed distribution maps.
- **Temporal Monitoring**: GIS allows for the comparison of weed maps over time, helping track the effectiveness of weed control measures and identifying recurring problem areas.
- **Example**: A study by Jha *et al.* (2021) demonstrated that GPS and GIS-based weed mapping reduced herbicide use by 20 per cent in precision agriculture systems by targeting only infested zones.

2. Site-Specific Weed Management

One of the most significant advantages of GPS/GIS and VRT is the ability to implement site-specific weed management (SSWM). This approach focuses on treating only the areas of the field where weeds are present, rather than applying uniform treatments.

- **Reduced Herbicide Use**: By targeting specific zones, farmers can significantly reduce the volume of herbicides applied, lowering input costs and minimizing environmental contamination.
- **Improved Efficacy**: Site-specific application ensures that weeds receive the appropriate type and dosage of herbicide, improving control efficacy and reducing the risk of herbicide resistance.

- ☆ **Example**: In a precision farming trial, VRT sprayers equipped with GPS reduced herbicide application rates by 30 per cent while maintaining effective weed control (Shaner, 2020).

3. Cost Savings

The adoption of GPS, GIS, and VRT in weed control leads to substantial cost savings for farmers by optimizing resource use.

- ☆ **Lower Input Costs**: Reduced herbicide usage directly translates to lower input costs. Additionally, targeted applications reduce fuel consumption and wear on machinery.
- ☆ **Labor Efficiency**: Automated weed detection and mapping reduce the need for manual scouting, saving labor costs and time.
- ☆ **Economic Analysis**: Studies have shown that precision weed management using VRT can reduce overall weed control costs by 15–25 per cent compared to conventional blanket applications (Shaner, 2020).

4. Environmental Sustainability

Precision weed management using GPS/GIS and VRT aligns with sustainable agriculture practices by minimizing the environmental footprint of weed control.

- ☆ **Reduced Chemical Runoff**: Targeted herbicide applications reduce the risk of chemical runoff into water bodies, protecting aquatic ecosystems.
- ☆ **Lower Carbon Emissions**: Optimized machinery use and reduced herbicide production contribute to lower greenhouse gas emissions.
- ☆ **Soil Health Preservation**: Avoiding over-application of herbicides helps maintain soil microbial diversity and overall soil health.

5. Real-Time Decision-Making

GPS and GIS technologies enable real-time decision-making, allowing farmers to respond quickly to weed infestations.

- ☆ **Dynamic Adjustments**: VRT sprayers equipped with GPS can adjust herbicide application rates in real-time based on weed density and type.
- ☆ **Integration with Sensors**: Advanced weed detection systems, such as optical sensors or drones, can feed real-time data into GIS platforms, enabling immediate action.
- ☆ **Example**: A farmer using real-time weed detection with VRT sprayers can treat a newly identified weed patch during the same field pass, saving time and resources.

6. Improved Data Management and Analysis

GIS platforms serve as powerful tools for managing and analyzing weed-related data, aiding in long-term weed control planning.

- **Centralized Data Storage**: GIS allows for the integration of various data sources, including weed maps, soil properties, and crop yield data, into a single platform.
- **Predictive Modeling**: Historical weed data can be used to predict future infestations and plan proactive control measures.
- **Collaboration and Knowledge Sharing**: GIS data can be shared with agronomists, researchers, and extension services to develop region-specific weed management strategies.

7. Integration with Other Precision Agriculture Tools

GPS/GIS and VRT technologies can be seamlessly integrated with other precision agriculture tools, enhancing overall farm management.

- **Drones and UAVs**: Unmanned aerial vehicles equipped with GPS and imaging sensors can provide high-resolution weed maps for GIS analysis.
- **Soil and Crop Sensors**: Data from soil moisture or nutrient sensors can be overlaid with weed maps in GIS to understand weed-crop interactions better.
- **Example**: Combining GPS-guided weed mapping with drone-based imagery allowed researchers to achieve 95 per cent accuracy in identifying weed species in a cornfield (Jha *et al.*, 2021).

8. Scalability and Adaptability

GPS, GIS, and VRT technologies are scalable and adaptable to various farm sizes and cropping systems.

- **Small and Large Farms**: While large farms benefit significantly from these technologies, smallholder farmers can also adopt scaled-down versions, such as handheld GPS devices and portable VRT sprayers.
- **Diverse Cropping Systems**: These technologies are applicable across different crops, from cereals to horticultural crops, making them versatile tools for weed management.

F. Challenges and Considerations for GPS/GIS and Variable Rate Technology in Weed Control

The integration of GPS (Global Positioning System), GIS (Geographic Information System), and Variable Rate Technology (VRT) in weed management has revolutionized modern agriculture. These technologies enable site-specific weed control, reducing herbicide usage, enhancing efficiency, and minimizing environmental impact. However, their adoption is not without challenges. This section explores the key challenges and considerations associated with the use of these advanced tools in weed control.

1. Technical Challenges

Data Accuracy and Precision

- GPS and GIS systems rely on accurate geospatial data to map weed infestations and guide variable rate applications. However:
- Signal interference from buildings, trees, or atmospheric conditions can reduce GPS accuracy, leading to errors in weed mapping or herbicide application.
- GIS data layers, such as soil type, topography, and weed distribution, may not always be up-to-date or detailed enough for precise decision-making.
- Differential GPS (DGPS) and Real-Time Kinematic (RTK) systems can improve accuracy, but they require additional infrastructure and investment.

Sensor Limitations

- Weed detection sensors, such as optical or multispectral sensors, may struggle to differentiate between crop plants and weeds in mixed stands or under dense canopies.
- Environmental factors, such as lighting conditions, soil moisture, or dust, can interfere with sensor performance.
- Emerging technologies like hyperspectral imaging and machine learning algorithms are promising but require further development and validation for field use.

Equipment Compatibility

- Integrating GPS/GIS systems with existing farm equipment can be challenging due to compatibility issues between hardware and software from different manufacturers.
- Standardization of data formats and communication protocols is essential to ensure seamless operation across diverse systems.

2. Economic Considerations

High Initial Investment

- The cost of acquiring GPS/GIS systems, VRT-enabled machinery, and associated software can be prohibitive for small and medium-sized farms.

- Additional expenses include training personnel, maintaining equipment, and subscribing to data services.

Cost-Benefit Analysis

- The economic benefits of GPS/GIS and VRT in weed control depend on factors such as weed pressure, field size, and crop value.
- In low-weed-density scenarios, the cost savings from reduced herbicide use may not justify the investment in advanced technologies.
- Farmers must conduct a detailed cost-benefit analysis to determine the economic feasibility of adopting these tools.

3. Operational Challenges

Complexity of Implementation

- Implementing GPS/GIS and VRT requires technical expertise in data collection, analysis, and interpretation.
- Farmers and operators need training to effectively use these systems, which can be time-consuming and resource-intensive.
- Errors in data input or system calibration can lead to suboptimal weed control or crop damage.

Scalability

- While GPS/GIS and VRT are highly effective in large-scale farming operations, their scalability to smallholder farms remains a challenge.
- Small fields with irregular shapes may require customized solutions, increasing complexity and costs.

Data Management

- Collecting, storing, and analyzing large volumes of geospatial and weed-related data can be overwhelming.
- Ensuring data security and privacy is another concern, particularly when using cloud-based platforms.

4. Environmental and Regulatory Considerations

Environmental Impact

- While VRT reduces herbicide use, improper calibration or application errors can lead to over-application in certain areas, causing environmental harm.
- The disposal of outdated or malfunctioning electronic components from GPS/GIS systems raises concerns about e-waste management.

Regulatory Compliance

- Farmers must ensure that GPS/GIS and VRT systems comply with local and international regulations governing pesticide application and data usage.
- Regulatory frameworks for precision agriculture technologies are still evolving, creating uncertainty for users.

5. Social and Knowledge Barriers

Knowledge Gaps

- Many farmers lack awareness or understanding of GPS/GIS and VRT technologies, limiting their adoption.
- Extension services and training programs are essential to bridge this knowledge gap and promote the benefits of precision weed management.

Resistance to Change

- Farmers accustomed to traditional weed control methods may be hesitant to adopt new technologies due to perceived risks or skepticism about their effectiveness.
- Demonstration projects and success stories can help build confidence and encourage adoption.

6. Future Directions and Opportunities

Integration with Artificial Intelligence (AI)

- AI-powered decision support systems can enhance the effectiveness of GPS/GIS and VRT by analyzing complex datasets and providing real-time recommendations.
- For example, machine learning algorithms can improve weed identification and optimize herbicide application rates.

Advancements in Connectivity

- The expansion of IoT (Internet of Things) and 5G networks can facilitate seamless communication between GPS/GIS systems, sensors, and machinery.
- Improved connectivity will enable real-time data sharing and remote monitoring of weed control operations.

Collaborative Efforts

- Collaboration between researchers, technology providers, and farmers is crucial to address the challenges of GPS/GIS and VRT in weed control.

- ☆ Public-private partnerships can support the development of affordable and user-friendly solutions tailored to diverse farming contexts.

G. Future Directions for GPS/GIS and Variable Rate Technology in Weed Control

The integration of Geographic Positioning Systems (GPS), Geographic Information Systems (GIS), and Variable Rate Technology (VRT) has revolutionized weed management by enabling site-specific applications and precision agriculture techniques. As these technologies continue to evolve, their future applications in weed control promise to enhance efficiency, reduce environmental impacts, and improve economic returns for farmers. This section explores the potential advancements and future directions in the use of GPS/GIS and VRT for weed management.

1. Enhanced Weed Mapping and Monitoring with Advanced GIS

GIS technology has already proven its utility in mapping weed infestations, but future advancements are expected to make weed mapping more accurate and dynamic. Emerging trends include:

- ☆ **Integration with Remote Sensing and Drones**: High-resolution satellite imagery and drone-mounted sensors are becoming increasingly accessible. These tools can collect real-time data on weed distribution and density, which can be integrated into GIS platforms for detailed weed maps. For example, hyperspectral and multispectral imaging can differentiate between crop species and weeds based on their spectral signatures, enabling precise identification of weed patches (Zhang *et al.*, 2022).
- ☆ **Machine Learning and Artificial Intelligence (AI)**: AI algorithms are being developed to analyze GIS data and identify weed patterns with greater accuracy. These algorithms can process large datasets from multiple sources, such as drones, satellites, and ground sensors, to predict weed outbreaks and assess weed pressure in real-time.
- ☆ **Dynamic Weed Maps**: Future GIS systems will likely incorporate temporal data, allowing for the creation of dynamic weed maps that track weed growth stages, seasonal variations, and herbicide resistance trends. This will enable farmers to adjust their weed management strategies throughout the growing season.

2. Advancements in Variable Rate Technology (VRT) for Weed Control

Variable Rate Technology (VRT) is a cornerstone of precision weed management, allowing for the targeted application of herbicides and other inputs. Future developments in VRT are expected to focus on the following areas:

- ☆ **Real-Time Weed Detection Systems**: Current VRT systems often rely on pre-mapped weed data, but future systems will incorporate real-time weed detection. For instance, camera-based sensors mounted on sprayers can identify weeds as they pass through the field and adjust herbicide application rates accordingly.
- ☆ **Integration with Autonomous Machinery**: Autonomous tractors and robotic weeders equipped with VRT capabilities are expected to become more prevalent. These machines can navigate fields using GPS guidance, detect weeds using advanced sensors, and apply herbicides or mechanical control measures with precision.
- ☆ **Multi-Layered Data Integration**: Future VRT systems will integrate data from multiple sources, such as soil nutrient maps, crop health data, and weed pressure maps, to optimize herbicide application rates. For example, areas with high weed density and low crop competition may receive higher herbicide doses, while areas with low weed pressure may receive minimal or no treatment.
- ☆ **Site-Specific Herbicide Formulations**: Advances in VRT will also enable the use of site-specific herbicide formulations. By combining GPS and GIS data with weed species identification, farmers can apply tailored herbicide mixtures that target specific weed species while minimizing off-target effects.

3. Improved Decision Support Systems (DSS) for Weed Management

Decision Support Systems (DSS) are software tools that help farmers make informed decisions about weed control strategies. The future of DSS in weed management will likely involve:

- ☆ **Integration with Big Data and IoT**: Internet of Things (IoT) devices, such as soil moisture sensors and weather stations, can provide real-time data that feeds into DSS platforms. This data can be combined with GPS and GIS information to recommend optimal weed control measures.
- ☆ **Predictive Weed Management Models**: DSS platforms will incorporate predictive models that use historical data, weather forecasts, and weed biology to anticipate weed outbreaks. These models can help farmers plan preemptive control measures, reducing the need for reactive herbicide applications.
- ☆ **User-Friendly Interfaces**: As DSS platforms become more sophisticated, efforts will be made to ensure they remain accessible to farmers. Mobile apps and cloud-based platforms will allow users to access DSS tools from anywhere, making precision weed management more practical for small-scale and resource-limited farmers.

4. Sustainability and Environmental Considerations

The future of GPS/GIS and VRT in weed control will also focus on sustainability and environmental stewardship. Key developments include:

- **Reduction in Herbicide Use**: Precision application techniques enabled by GPS and VRT can significantly reduce herbicide use, lowering the risk of environmental contamination and herbicide resistance development.
- **Carbon Footprint Reduction**: By optimizing field operations and minimizing unnecessary chemical applications, these technologies can reduce the carbon footprint of weed management practices.
- **Integration with Organic Farming Practices**: GPS and GIS technologies can also support non-chemical weed control methods, such as mechanical weeding and cover cropping, by identifying areas where these practices are most effective.

5. Challenges and Opportunities

While the future of GPS/GIS and VRT in weed control is promising, several challenges remain:

- **Data Management and Standardization**: The integration of data from multiple sources requires standardized formats and protocols to ensure compatibility and usability.
- **Cost and Accessibility**: Advanced GPS/GIS and VRT systems can be expensive, limiting their adoption by small-scale farmers. Efforts to reduce costs and develop scalable solutions will be critical.
- **Training and Education**: Farmers and agronomists will need training to effectively use these technologies and interpret the data they generate.

Despite these challenges, the continued development of GPS/GIS and VRT technologies offers immense potential for transforming weed management practices, making them more efficient, sustainable, and economically viable.

Chapter 6

Drones and Remote Sensing

The integration of drones and remote sensing technologies in modern weed management has revolutionized precision agriculture. These tools provide accurate, real-time data for identifying, monitoring, and managing weed infestations with minimal environmental impact.

Applications of Drones in Weed Management

- **Weed Mapping**: Drones equipped with multispectral and hyperspectral cameras can detect weed infestations by analyzing differences in vegetation indices such as the Normalized Difference Vegetation Index (NDVI). This enables precise mapping of weed distribution across fields.
- **Targeted Herbicide Application**: By pinpointing weed hotspots, drones can guide sprayers for site-specific herbicide application, reducing chemical use and costs.
- **Monitoring Herbicide Efficacy**: Post-application drone surveys help assess the effectiveness of weed control measures, allowing timely adjustments.

Remote Sensing Technologies

- **Multispectral Imaging**: Captures data in specific wavelength bands to distinguish weeds from crops based on spectral signatures.
- **Thermal Imaging**: Detects temperature variations, which can indicate stress caused by weed competition.
- **LiDAR (Light Detection and Ranging)**: Provides 3D mapping of vegetation, enabling precise weed height and density assessments.

Advantages

- ☆ Enhanced accuracy in weed detection.
- ☆ Reduced labor and operational costs.
- ☆ Minimized environmental impact through precise chemical application.

Challenges

- ☆ High initial investment in equipment.
- ☆ Need for technical expertise in data interpretation.
- ☆ Limitations in detecting weeds under dense crop canopies.

Emerging advancements, such as machine learning algorithms integrated with drone data, are further improving weed identification and management strategies (Zhang *et al.*, 2022). These technologies hold immense potential for sustainable and efficient weed control.

A. Introduction to Drones and Remote Sensing

The integration of drones and remote sensing technologies into agriculture has revolutionized modern weed management practices. These advanced tools enable precise, efficient, and cost-effective monitoring and control of weeds, significantly reducing the reliance on traditional, labor-intensive methods. This section explores the fundamentals of drones and remote sensing, their applications in weed management, and the underlying scientific principles.

Drones: A Game-Changer in Agriculture

Drones, also known as unmanned aerial vehicles (UAVs), are aircraft systems operated remotely or autonomously. Equipped with advanced sensors and cameras, drones provide high-resolution imagery and real-time data, making them invaluable for agricultural applications. Their ability to cover large areas quickly and repeatedly has made them a cornerstone of precision agriculture.

Key features of drones in agriculture include:

- ☆ **High-Resolution Imaging:** Drones can capture images with resolutions as fine as a few centimeters per pixel, enabling detailed analysis of weed infestations.
- ☆ **Flexibility and Accessibility:** Drones can operate in diverse terrains and weather conditions, providing access to areas that are difficult to reach by traditional means.
- ☆ **Cost-Effectiveness:** Compared to satellite imagery, drones offer a more affordable and customizable solution for small and medium-scale farms.

In weed management, drones are primarily used for weed mapping, herbicide application, and monitoring the effectiveness of control measures.

Remote Sensing: The Science Behind Precision

Remote sensing refers to the process of collecting information about an object or area without direct physical contact. In agriculture, this is achieved using sensors mounted on drones, satellites, or ground-based platforms. These sensors detect and measure electromagnetic radiation reflected or emitted by plants, soil, and other surfaces.

Types of Remote Sensing Sensors

1. Multispectral Sensors

These sensors capture data in multiple wavelengths, including visible (red, green, blue) and near-infrared (NIR) bands. Multispectral imagery is widely used to differentiate between crops and weeds based on their spectral signatures.

2. Hyperspectral Sensors

Hyperspectral sensors capture data in hundreds of narrow spectral bands, providing detailed information about plant health, stress, and species identification. Although more expensive, they offer unparalleled accuracy in weed detection.

3. Thermal Sensors

Thermal sensors measure temperature variations, which can indicate water stress or differences in vegetation types, aiding in weed identification.

4. LiDAR (Light Detection and Ranging)

LiDAR uses laser pulses to create 3D maps of the terrain, helping identify weed-infested areas in complex landscapes.

Applications in Weed Management

1. Weed Mapping and Identification

Drones equipped with multispectral or hyperspectral sensors can distinguish weeds from crops by analyzing their spectral reflectance patterns. For instance, weeds often exhibit higher reflectance in the red and near-infrared bands compared to crops. This data is processed to create weed distribution maps, enabling targeted interventions.

2. Precision Herbicide Application

Drones can be integrated with variable-rate sprayers to apply herbicides precisely to weed-infested areas, minimizing chemical usage and environmental impact. This method, known as site-specific weed management (SSWM), reduces costs and enhances sustainability (Figure 6.1).

Figure 6.1. Efficient Agricultural Drone.

3. Monitoring Weed Control Effectiveness

Post-treatment monitoring is critical to evaluate the success of weed management strategies. Drones provide time-series data, allowing farmers to assess changes in weed density and distribution over time.

4. Early Detection of Weed Infestations

Remote sensing enables the early detection of weed outbreaks, allowing timely interventions before the infestation spreads. This is particularly useful for managing invasive weed species.

Scientific Principles Underpinning Remote Sensing

The effectiveness of remote sensing in weed management relies on the concept of spectral signatures. Every plant species reflects and absorbs light differently across various wavelengths due to differences in leaf structure, pigment composition, and water content. By analyzing these spectral patterns, remote sensing technologies can differentiate between crops and weeds with high accuracy.

The Normalized Difference Vegetation Index (NDVI) is a commonly used metric in remote sensing. NDVI values are calculated using the formula:

$$NDVI = (NIR - Red)/(NIR + Red)$$

Where NIR represents near-infrared reflectance and Red represents red reflectance. Healthy vegetation typically exhibits high NDVI values, while stressed or weedy areas show lower values.

Recent Advances and Future Prospects

Recent advancements in machine learning and artificial intelligence (AI) have further enhanced the capabilities of drones and remote sensing. Algorithms can now process large datasets to classify weeds, predict their growth patterns, and recommend optimal control measures. Additionally, the advent of autonomous drones equipped with AI-driven navigation systems promises to make weed management even more efficient.

According to a study by Zhang *et al.* (2022a), the integration of AI with drone-based remote sensing improved weed detection accuracy by up to 95 per cent, highlighting the potential of these technologies in modern agriculture.

B. Applications of Drones in Weed Control

The integration of drones, also known as Unmanned Aerial Vehicles (UAVs), into agricultural practices has revolutionized weed management. These advanced tools offer precision, efficiency, and cost-effectiveness, making them indispensable in modern weed control strategies. By leveraging cutting-edge technologies such as remote sensing, artificial intelligence, and variable rate application systems, drones enable farmers and researchers to monitor, identify, and manage weeds with unparalleled accuracy.

1. Weed Mapping and Monitoring

Drones equipped with high-resolution cameras and multispectral or hyperspectral sensors provide detailed aerial imagery of agricultural fields. These images are processed using specialized software to detect and map weed infestations.

- ☆ **Multispectral and Hyperspectral Imaging**: These sensors capture data across various wavelengths, allowing differentiation between crops and weeds based on their spectral signatures. For instance, weeds often exhibit distinct reflectance patterns in the near-infrared spectrum compared to crops.
- ☆ **Real-Time Monitoring**: Drones can survey large areas in a fraction of the time required by manual scouting. This enables real-time monitoring of weed growth and distribution, allowing timely interventions.
- ☆ **Data Integration with GIS**: Weed maps generated by drones can be integrated with Geographic Information Systems (GIS) to create spatial weed distribution models. This aids in understanding weed dynamics and planning targeted control measures.

2. Precision Herbicide Application

One of the most significant applications of drones in weed control is their ability to perform site-specific herbicide application. This reduces chemical usage, minimizes environmental impact, and lowers costs.

- ☆ **Spot Spraying**: Drones equipped with sprayer systems can apply herbicides directly to weed-infested areas, avoiding unnecessary treatment of weed-free zones. This is particularly useful in fields with patchy weed distribution.
- ☆ **Variable Rate Application**: Advanced drones use algorithms to adjust the herbicide dosage based on weed density and type. For example, denser weed patches may require higher herbicide concentrations, while sparser areas need less.
- ☆ **Minimized Drift**: Drones fly close to the ground and use precision nozzles, reducing herbicide drift and ensuring accurate application.

3. Early Detection of Herbicide Resistance

Drones play a crucial role in identifying herbicide-resistant weed populations. By capturing detailed imagery and analyzing plant health indicators, drones can detect subtle differences in weed response to herbicides.

- ☆ **Stress Detection**: Multispectral imaging can identify stressed weeds that survive herbicide application, indicating potential resistance. Early detection allows farmers to adjust their weed management strategies and prevent the spread of resistant populations.
- ☆ **Data Analytics**: Machine learning algorithms process drone-collected data to classify weeds based on their response to herbicides, providing insights into resistance patterns.

4. Cost and Labor Efficiency

Drones significantly reduce the labor and time required for weed management tasks. Traditional methods such as manual scouting and blanket herbicide application are labor-intensive and often less precise. Drones, on the other hand, can cover hundreds of hectares in a single flight, providing detailed data and performing targeted interventions.

- ☆ **Reduced Input Costs**: By optimizing herbicide usage, drones lower input costs while maintaining effective weed control.
- ☆ **Labor Savings**: The automation of weed scouting and spraying reduces the dependency on manual labor, addressing labor shortages in agriculture.

5. Environmental Benefits

The precision offered by drones contributes to sustainable weed management practices. By minimizing herbicide use, drones reduce the risk of chemical runoff into water bodies and soil contamination. Additionally, targeted application protects non-target organisms, including beneficial insects and neighboring crops.

- **Carbon Footprint Reduction**: Drones consume less fuel compared to traditional machinery, contributing to a lower carbon footprint.
- **Biodiversity Conservation**: Precision spraying ensures that only weeds are targeted, preserving the surrounding ecosystem.

6. Challenges and Future Prospects

While drones offer immense potential, their adoption in weed control faces certain challenges, including high initial costs, regulatory restrictions, and the need for technical expertise. However, advancements in drone technology, such as increased payload capacity, longer flight times, and improved AI algorithms, are expected to address these limitations.

- **Integration with AI and IoT**: Future drones will likely integrate with Internet of Things (IoT) devices and artificial intelligence to provide real-time, autonomous weed management solutions.
- **Scalability**: As drone technology becomes more affordable, its adoption is expected to increase, particularly among smallholder farmers

C. Aerial Weed Mapping

Aerial weed mapping is an advanced precision agriculture technique that utilizes aerial imagery to identify, monitor, and manage weed infestations in agricultural fields. This method leverages cutting-edge technologies such as drones, satellites, and manned aircraft equipped with high-resolution cameras and sensors to collect data on weed distribution. By providing spatially accurate and timely information, aerial weed mapping enables farmers and agronomists to implement targeted weed control strategies, reducing herbicide use and improving crop productivity.

Technologies Used in Aerial Weed Mapping

1. Unmanned Aerial Vehicles (UAVs) or Drones

- Drones are widely used for aerial weed mapping due to their affordability, ease of deployment, and ability to capture high-resolution images.
- Equipped with multispectral or hyperspectral sensors, drones can differentiate between crops and weeds based on their spectral signatures.
- For example, weeds such as Palmer amaranth (Amaranthus palmeri) exhibit distinct reflectance patterns in the near-infrared spectrum, which can be detected by drone-mounted sensors.

2. **Satellite Imagery**

- Satellites provide large-scale imagery, making them suitable for mapping weeds in extensive agricultural landscapes.
- Modern satellites like Sentinel-2 and Landsat 8 offer multispectral and temporal data, enabling the identification of weed patches over time.
- However, satellite imagery may lack the spatial resolution required for small-scale weed detection.

3. **Manned Aircraft**

- Fixed-wing aircraft or helicopters equipped with advanced imaging systems can cover vast areas quickly, making them ideal for large-scale weed mapping.
- These systems often include LiDAR (Light Detection and Ranging) and thermal imaging sensors for enhanced weed detection in complex terrains.

Data Collection and Processing

1. **Image Acquisition**

- Aerial platforms capture images of the field, which are georeferenced using GPS technology to ensure spatial accuracy.
- Multispectral and hyperspectral imaging are commonly employed to capture data across various wavelengths, enabling the differentiation of plant species.

2. **Data Preprocessing**

- Raw imagery is processed to remove noise, correct distortions, and enhance image quality.
- Techniques such as radiometric calibration and orthorectification are applied to ensure accurate analysis.

3. **Weed Identification**

- Advanced algorithms, including machine learning and artificial intelligence (AI), are used to classify vegetation based on spectral and spatial features.
- For instance, convolutional neural networks (CNNs) have demonstrated high accuracy in distinguishing weeds from crops in aerial imagery (Lottes *et al.*, 2017).

4. **Mapping and Visualization**

- The processed data is used to create weed distribution maps, which can be integrated into Geographic Information Systems (GIS) for further analysis.

- These maps provide actionable insights, such as the location, density, and type of weed infestations.

Applications of Aerial Weed Mapping

- **Site-Specific Weed Management (SSWM):** Aerial weed maps enable farmers to apply herbicides only where weeds are present, reducing chemical usage and minimizing environmental impact.
- **Weed Monitoring Over Time:** Temporal data from aerial imagery can track weed growth patterns and assess the effectiveness of weed control measures.
- **Early Detection of Invasive Weeds:** Aerial mapping facilitates the early identification of invasive weed species, allowing for timely intervention and containment.
- **Integration with Variable Rate Technology (VRT):** Weed maps can be integrated with VRT systems to automate herbicide application, further enhancing precision and efficiency.

Advantages of Aerial Weed Mapping

- **High Efficiency:** Aerial platforms can cover large areas in a short time, making them ideal for extensive agricultural fields.
- **Improved Accuracy:** Advanced imaging technologies and AI-based analysis ensure precise weed identification and mapping.
- **Cost-Effectiveness:** While initial investment in drones or sensors may be high, the long-term savings from reduced herbicide use and improved crop yields outweigh the costs.
- **Environmental Benefits:** Targeted weed control minimizes chemical runoff and reduces the risk of herbicide resistance development.

Challenges and Limitations

- **High Initial Costs:** The acquisition of drones, sensors, and software can be expensive, particularly for small-scale farmers.
- **Data Processing Complexity:** Analyzing aerial imagery requires technical expertise and access to advanced software tools.
- **Weather Dependency:** Cloud cover, wind, and other weather conditions can affect image quality and data collection.
- **Resolution Trade-Offs:** While drones provide high-resolution data, their coverage area is limited compared to satellites.

Future Prospects

The integration of aerial weed mapping with emerging technologies such as Internet of Things (IoT), blockchain, and autonomous farming systems holds immense potential. For instance, real-time weed detection using drone swarms and AI-driven decision-making could revolutionize weed management practices.

D. Targeted Herbicide Spraying using Drones and Remote Sensing

The integration of drones and remote sensing technologies into modern agriculture has revolutionized weed management practices. Targeted herbicide spraying, enabled by these advancements, offers precision, efficiency, and sustainability. This section explores the principles, technologies, and applications of targeted herbicide spraying using drones and remote sensing.

Principles of Targeted Herbicide Spraying

Targeted herbicide spraying involves the precise application of herbicides to specific areas infested with weeds, minimizing chemical use and reducing environmental impact. This approach relies on advanced technologies to detect, map, and treat weed infestations with high accuracy. The key principles include:

- ☆ **Precision Agriculture**: Utilizing data-driven techniques to optimize inputs and maximize outputs.
- ☆ **Site-Specific Weed Management (SSWM)**: Focusing on localized weed control rather than blanket applications.
- ☆ **Minimization of Non-Target Effects**: Reducing herbicide exposure to crops, soil, and non-target organisms.

Role of Drones in Weed Management

Drones, also known as unmanned aerial vehicles (UAVs), have become indispensable tools in modern weed management. They offer several advantages over traditional methods:

- ☆ **Rapid Data Collection**: Drones can survey large fields in a short time, capturing high-resolution images for weed detection.
- ☆ **Accessibility**: They can operate in challenging terrains where ground-based equipment may struggle.
- ☆ **Cost-Effectiveness**: Reduced labor and operational costs compared to manual or tractor-based spraying.
- ☆ **Precision Spraying**: Equipped with advanced nozzles, drones can deliver herbicides directly to weed-infested areas.

Components of a Drone-Based Weed Management System

- ☆ **Imaging Sensors**: Multispectral, hyperspectral, and RGB cameras capture data for weed identification.
- ☆ **GPS and GIS Integration**: Enables accurate geolocation and mapping of weed patches.
- ☆ **Spraying Mechanism**: Drones are equipped with variable-rate sprayers to apply herbicides precisely.
- ☆ **Software and Algorithms**: Machine learning algorithms analyze images to distinguish weeds from crops.

Remote Sensing for Weed Detection

Remote sensing is the backbone of targeted herbicide spraying, providing the data needed to identify and map weed infestations. It involves the use of sensors mounted on drones, satellites, or other platforms to collect information about the field. Key aspects include:

Types of Sensors Used

- ☆ **Multispectral Sensors**: Capture data in multiple wavelengths, enabling the differentiation of vegetation types based on their spectral signatures.
- ☆ **Hyperspectral Sensors**: Provide detailed spectral information, allowing for the identification of specific weed species.
- ☆ **Thermal Sensors**: Detect temperature variations, which can indicate weed stress or growth patterns.

Weed Identification Techniques

- ☆ **Normalized Difference Vegetation Index (NDVI)**: Measures vegetation health and density, helping to distinguish weeds from crops.
- ☆ **Machine Learning Models**: Algorithms such as convolutional neural networks (CNNs) are trained to classify weeds based on image data.
- ☆ **Spectral Signature Analysis**: Identifies weeds based on their unique reflectance patterns in different wavelengths.

Advantages of Remote Sensing

- ☆ **Non-Invasive**: Data collection does not disturb the field or crops.
- ☆ **High Accuracy**: Advanced sensors and algorithms ensure precise weed identification.
- ☆ **Scalability**: Suitable for small farms as well as large agricultural operations.

Applications of Targeted Herbicide Spraying

The combination of drones and remote sensing has numerous applications in weed management:

1. **Early Detection of Weeds**: Remote sensing enables the identification of weed infestations at an early stage, allowing for timely intervention.
2. **Variable-Rate Herbicide Application**: Drones can adjust the amount of herbicide applied based on weed density, reducing chemical use.
3. **Integrated Weed Management (IWM)**: Data from drones and sensors can be integrated with other weed control methods for a holistic approach.
4. **Monitoring and Evaluation**: Post-spraying surveys using drones can assess the effectiveness of herbicide application.

Challenges and Future Prospects

Despite its potential, targeted herbicide spraying using drones and remote sensing faces several challenges:

- **High Initial Costs**: The equipment and software required can be expensive for small-scale farmers.
- **Regulatory Restrictions**: Drone usage is subject to strict regulations in many countries.
- **Technical Expertise**: Farmers need training to operate drones and interpret remote sensing data.

Future advancements in sensor technology, machine learning, and drone capabilities are expected to address these challenges. For instance, the development of lightweight hyperspectral sensors and autonomous drones will further enhance the efficiency and accessibility of this technology.

Case Studies and Research Insights

Recent studies highlight the effectiveness of targeted herbicide spraying:

- A study by Peña *et al.* (2023) demonstrated that drone-based spraying reduced herbicide use by 30 per cent while maintaining weed control efficacy in soybean fields.
- Research by Zhang *et al.* (2022b) showed that integrating hyperspectral imaging with machine learning improved weed species classification accuracy to over 90 per cent.

These findings underscore the potential of drones and remote sensing to transform weed management practices.

E. Monitoring and Data Collection using Drones and Remote Sensing

The integration of drones and remote sensing technologies has revolutionized weed management by enabling precise monitoring and data collection. These tools provide real-time, high-resolution data, allowing agricultural professionals to identify weed infestations, assess their severity, and implement targeted control measures. This section explores the principles, applications, and benefits of drones and remote sensing in modern weed management.

Principles of Drones and Remote Sensing in Weed Monitoring

Drones, also known as unmanned aerial vehicles (UAVs), are equipped with advanced sensors and cameras that capture detailed imagery of agricultural fields. Remote sensing refers to the collection of data from a distance, typically using satellite or aerial platforms, to analyze crop health, soil conditions, and weed distribution. These technologies operate on the principles of spectral imaging and spatial analysis:

- ☆ **Spectral Imaging**: Vegetation reflects light differently across the electromagnetic spectrum. Healthy crops and weeds exhibit distinct spectral signatures, particularly in the visible (VIS) and near-infrared (NIR) regions. Multispectral and hyperspectral sensors on drones and satellites capture these variations, enabling the differentiation of weeds from crops.
- ☆ **Spatial Analysis**: High-resolution imagery allows for the precise mapping of weed infestations. By analyzing spatial patterns, professionals can identify areas of concern and prioritize management efforts.

Applications in Weed Management

Drones and remote sensing technologies have diverse applications in weed monitoring and data collection:

1. Weed Identification and Mapping

- ☆ Drones equipped with multispectral cameras can detect weed species based on their unique spectral signatures. For example, broadleaf weeds and grasses exhibit different reflectance patterns, enabling accurate identification.
- ☆ Geographic Information System (GIS) software processes drone imagery to create detailed weed distribution maps, which guide site-specific management strategies.

2. Weed Density Assessment

- Remote sensing provides quantitative data on weed density and biomass. This information is crucial for determining the severity of infestations and selecting appropriate control measures.
- For instance, Normalized Difference Vegetation Index (NDVI) maps generated from drone data can highlight areas with high weed pressure.

3. Monitoring Herbicide Efficacy

- Post-application monitoring is essential to evaluate the effectiveness of herbicide treatments. Drones can capture time-series images to track changes in weed cover and assess treatment success.
- This approach reduces reliance on manual field scouting, saving time and labor.

4. Early Detection of Weed Infestations

- Early detection is critical for preventing the spread of invasive weed species. Drones can survey large areas quickly, identifying emerging infestations before they become unmanageable.

Advantages of Using Drones and Remote Sensing

The adoption of drones and remote sensing in weed management offers several advantages:

- **Precision**: High-resolution imagery enables the identification of individual weeds, supporting targeted interventions and reducing the overuse of herbicides.
- **Efficiency**: Drones can cover large areas in a short time, significantly reducing the time and labor required for field scouting.
- **Cost-Effectiveness**: While the initial investment in drones and sensors may be high, the long-term savings in labor, herbicide use, and crop losses outweigh the costs.
- **Environmental Sustainability**: By enabling precise application of herbicides, these technologies minimize chemical runoff and environmental contamination.

Challenges and Considerations

Despite their benefits, the use of drones and remote sensing in weed management comes with challenges:

- **Data Processing and Analysis**: The large volume of data generated by drones requires advanced software and expertise for processing and interpretation.

- **Weather Dependence**: Cloud cover, wind, and other weather conditions can affect the quality of drone imagery and remote sensing data.
- **Regulatory Compliance**: Operators must adhere to local regulations governing drone usage, including flight restrictions and licensing requirements.

Recent Advances and Future Prospects

Recent advancements in drone technology and remote sensing are enhancing their utility in weed management:

- **Artificial Intelligence (AI) and Machine Learning**: AI algorithms are being developed to automate weed identification and classification, improving accuracy and efficiency.
- **Integration with IoT**: Internet of Things (IoT) devices, such as soil moisture sensors, can be integrated with drone data to provide a comprehensive understanding of field conditions.
- **Miniaturization of Sensors**: The development of lightweight, high-resolution sensors is expanding the capabilities of drones for weed monitoring.

Case Study Example

A study conducted by Peña *et al.* (2023) demonstrated the effectiveness of drones in detecting Palmer amaranth (Amaranthus palmeri) in soybean fields. Using multispectral imaging, the researchers achieved an accuracy of over 90 per cent in identifying weed patches, highlighting the potential of this technology for site-specific weed management.

F. Role of Remote Sensing in Weed Control

Remote sensing has emerged as a transformative technology in modern agriculture, offering precise, efficient, and cost-effective solutions for weed management. By utilizing sensors, satellites, and drones, remote sensing enables the detection, monitoring, and mapping of weed infestations over large areas. This chapter explores the role of remote sensing in weed control, focusing on its principles, applications, and benefits for agricultural professionals.

Principles of Remote Sensing in Weed Management

Remote sensing involves the acquisition of data about the Earth's surface without direct contact. This is achieved through sensors mounted on satellites, drones, or aircraft, which capture electromagnetic radiation reflected or emitted by objects. In weed management, remote sensing relies on the following principles:

- **Spectral Signatures**: Different plant species, including weeds and crops, reflect and absorb light differently across various wavelengths. These unique spectral signatures allow for the differentiation between weeds and crops.
- **Spatial Resolution**: High-resolution imagery enables the identification of small weed patches, while lower-resolution imagery is suitable for large-scale monitoring.
- **Temporal Resolution**: Frequent data acquisition allows for the monitoring of weed growth and spread over time, aiding in timely interventions.

Applications of Remote Sensing in Weed Control

Remote sensing technologies have a wide range of applications in weed management, each contributing to improved decision-making and resource efficiency.

1. Weed Detection and Mapping

- **Multispectral and Hyperspectral Imaging**: Multispectral sensors capture data in a few broad wavelength bands, while hyperspectral sensors provide detailed information across hundreds of narrow bands. These technologies are used to identify weed species based on their spectral signatures.
- **Example**: Hyperspectral imaging has been successfully used to distinguish between invasive weed species like *Parthenium hysterophorus* and native vegetation in rangelands.
- **Normalized Difference Vegetation Index (NDVI)**: NDVI is a commonly used vegetation index derived from remote sensing data. It helps differentiate between healthy crops and weed-infested areas based on vegetation vigor.

2. Monitoring Weed Dynamics

- Remote sensing enables the tracking of weed growth stages, density, and spatial distribution over time. This information is critical for understanding weed-crop competition and planning targeted interventions.
- **Example**: Temporal satellite imagery has been used to monitor the spread of *Amaranthus palmeri* (Palmer amaranth) in soybean fields, allowing for timely herbicide application.

3. Precision Herbicide Application

- By integrating remote sensing data with GPS and variable-rate technology (VRT), farmers can apply herbicides only to weed-infested areas, reducing chemical use and environmental impact.

- **Example**: Drones equipped with multispectral cameras can identify weed patches and guide sprayers for site-specific herbicide application.

4. Early Weed Detection

- Early detection of weeds is crucial for preventing their establishment and spread. Remote sensing allows for the identification of weeds at their initial growth stages, enabling proactive management.
- **Example**: UAV-based remote sensing has been used to detect early infestations of *Avena fatua* (wild oats) in wheat fields.

5. Invasive Weed Management

- Remote sensing is particularly useful for managing invasive weed species in natural ecosystems and agricultural landscapes. It helps in mapping the extent of invasion and monitoring the effectiveness of control measures.
- **Example**: Satellite imagery has been used to map the invasion of *Prosopis juliflora* in arid regions, aiding in its management.

Technologies Used in Remote Sensing for Weed Control

Several remote sensing platforms and technologies are employed in weed management. These include:

1. Satellite-Based Remote Sensing

- Satellites like Sentinel-2 and Landsat provide multispectral imagery with high temporal resolution, making them suitable for large-scale weed monitoring.
- **Advantages**: Wide coverage, cost-effectiveness for large areas.
- **Limitations**: Lower spatial resolution compared to UAVs, weather dependency.

2. UAV (Drone)-Based Remote Sensing

- Drones equipped with multispectral or hyperspectral cameras offer high-resolution imagery and flexibility in data collection.
- **Advantages**: High spatial resolution, ability to fly below cloud cover.
- **Limitations**: Limited coverage area, higher operational costs.

3. Ground-Based Sensors

- Ground-based sensors, such as handheld spectrometers and proximal sensors, are used for detailed weed identification in small areas.
- **Advantages**: High accuracy, real-time data collection.
- **Limitations**: Limited to small-scale applications.

Benefits of Remote Sensing in Weed Control

Remote sensing offers numerous advantages for weed management, including:

- **Increased Efficiency**: Enables rapid assessment of large areas, saving time and labor.
- **Cost-Effectiveness**: Reduces input costs by minimizing unnecessary herbicide applications.
- **Environmental Sustainability**: Promotes site-specific management, reducing chemical runoff and environmental contamination.
- **Improved Decision-Making**: Provides accurate and timely data for informed weed control strategies.

Challenges and Limitations

Despite its potential, remote sensing in weed management faces certain challenges:

- **Data Interpretation**: Differentiating between weeds and crops with similar spectral signatures can be complex.
- **Cost of Technology**: High-resolution sensors and UAVs can be expensive for small-scale farmers.
- **Weather Dependency**: Cloud cover and adverse weather conditions can affect data quality, especially for satellite-based systems.

Future Prospects

Advancements in remote sensing technologies, such as the integration of artificial intelligence (AI) and machine learning (ML), are expected to enhance weed detection and classification. Additionally, the development of cost-effective sensors and platforms will make remote sensing accessible to a broader range of farmers.

G. Types of Sensors Used in Remote Sensing

Remote sensing has revolutionized modern weed management by enabling the precise monitoring of weed infestations across agricultural landscapes. Sensors play a pivotal role in capturing data that can be analyzed to identify weed species, density, and spatial distribution. These sensors are classified based on their operational principles, spectral sensitivity, and platforms. Below is an overview of the major types of sensors used in remote sensing for weed management.

1. Optical Sensors

Optical sensors are among the most widely used in remote sensing due to their ability to capture data in the visible, near-infrared (NIR), and shortwave infrared (SWIR) regions of the electromagnetic spectrum. These sensors detect reflected sunlight from vegetation and soil, which is then analyzed to differentiate between crops and weeds.

- **Multispectral Sensors**: These sensors capture data across a limited number of discrete spectral bands. They are commonly used to calculate vegetation indices like the Normalized Difference Vegetation Index (NDVI), which helps distinguish healthy crops from weeds.

  Example: Sensors on satellites like Sentinel-2 and Landsat-8 are equipped with multispectral capabilities.

- **Hyperspectral Sensors**: Hyperspectral sensors capture data across hundreds of narrow, contiguous spectral bands, providing detailed spectral signatures for vegetation. This allows for precise weed species identification based on their unique spectral reflectance patterns.

  Example: The AVIRIS (Airborne Visible/Infrared Imaging Spectrometer) sensor is a notable hyperspectral sensor used in agricultural research.

In weed management, optical sensors are particularly effective for detecting weeds with distinct spectral characteristics compared to crops. However, their reliance on sunlight makes them less effective under cloudy conditions or at night (Zhang *et al.*, 2022c).

2. Thermal Infrared Sensors

Thermal infrared sensors measure the thermal radiation emitted by objects, providing information about surface temperatures. In agriculture, these sensors can detect temperature differences between crops and weeds, especially under water stress conditions.

- **Application in Weed Management**: Weeds often exhibit different transpiration rates compared to crops, leading to variations in canopy temperature. Thermal sensors can exploit these differences to identify weed patches.
- **Platforms**: Thermal sensors are deployed on drones, aircraft, or satellites. For instance, the ECOSTRESS sensor aboard the International Space Station provides high-resolution thermal data for agricultural applications.

While thermal sensors are valuable for weed detection, their effectiveness can be influenced by environmental factors such as soil moisture and atmospheric conditions.

3. LiDAR Sensors

LiDAR (Light Detection and Ranging) sensors use laser pulses to measure distances and create high-resolution, three-dimensional maps of the Earth's surface. In agriculture, LiDAR is used to analyze crop canopy structure and detect variations caused by weed growth.

Key Features

- Provides detailed information on plant height and density.
- Effective in distinguishing weeds from crops based on structural differences.
- **Example**: LiDAR-equipped drones are increasingly used for mapping weed infestations in fields with complex topography.

  LiDAR's ability to operate in low-light conditions and its independence from sunlight make it a versatile tool for weed management.

4. Radar Sensors

Radar (Radio Detection and Ranging) sensors operate in the microwave region of the electromagnetic spectrum and are capable of penetrating clouds, smoke, and vegetation canopies. This makes them highly reliable for all-weather remote sensing.

- **Synthetic Aperture Radar (SAR)**: SAR sensors, such as those on the Sentinel-1 satellite, are widely used for agricultural monitoring. They can detect differences in surface roughness and moisture content, which are indicative of weed presence.

Advantages

- Operates day and night, regardless of weather conditions.
- Suitable for large-scale weed monitoring.

Radar sensors are particularly useful in regions with frequent cloud cover, where optical sensors may be limited.

5. Fluorescence Sensors

Fluorescence sensors detect the natural fluorescence emitted by plants during photosynthesis. This provides insights into plant health and stress levels, which can help differentiate between crops and weeds.

- **Chlorophyll Fluorescence**: Weeds and crops often exhibit distinct fluorescence patterns due to differences in chlorophyll content and photosynthetic activity.

Applications

Fluorescence sensors are used in precision agriculture to monitor weed stress and evaluate herbicide efficacy.

These sensors are typically mounted on UAVs (Unmanned Aerial Vehicles) for field-scale weed detection.

6. UAV-Mounted Sensors

Unmanned Aerial Vehicles (UAVs), or drones, have become a popular platform for deploying various sensors, including optical, thermal, and LiDAR sensors. UAV-mounted sensors offer high spatial resolution and flexibility in data acquisition.

Advantages

- Cost-effective for small to medium-sized farms.
- Provides real-time data for targeted weed control.

Example

UAVs equipped with multispectral cameras are widely used for mapping weed infestations and guiding site-specific herbicide applications.

7. Emerging Sensor Technologies

Recent advancements in sensor technology are paving the way for more precise and efficient weed management.

- **Hyperspectral LiDAR**: Combines the spectral capabilities of hyperspectral sensors with the structural mapping capabilities of LiDAR.
- **Quantum Sensors**: These sensors measure quantum properties such as photon entanglement and are being explored for advanced agricultural applications.

Emerging technologies hold promise for improving the accuracy and efficiency of weed detection in complex agricultural systems (Smith *et al.*, 2021).

H. Integration of Drones with AI and Machine Learning

The integration of drones with artificial intelligence (AI) and machine learning (ML) is revolutionizing modern weed management practices, offering precision, efficiency, and scalability. These technologies enable real-time weed detection, mapping, and targeted herbicide application, reducing labor costs and minimizing environmental impact. This section explores the applications, benefits, and underlying principles of using drones equipped with AI and ML in weed management.

Role of Drones in Weed Management

Drones, also known as unmanned aerial vehicles (UAVs), have become indispensable tools in precision agriculture. Their ability to cover large areas quickly and collect high-resolution data makes them ideal for weed management. When integrated with AI and ML, drones can perform the following tasks:

- **Weed Detection and Identification**: Drones equipped with multispectral or hyperspectral cameras capture high-resolution images of fields. AI algorithms analyze these images to differentiate between crops and weeds based on spectral signatures, shape, and texture.

- ☆ **Weed Mapping**: Using GPS and AI-driven image processing, drones create detailed weed distribution maps. These maps help farmers identify weed-infested zones and plan targeted interventions.
- ☆ **Spot Spraying**: Drones can be equipped with precision spraying systems that apply herbicides only to weed-infested areas, reducing chemical usage and protecting non-target plants.

AI and Machine Learning in Weed Management

AI and ML are the backbone of advanced weed management systems. These technologies enable drones to process vast amounts of data, learn from patterns, and make informed decisions. Key components include:

1. Image Recognition and Classification

AI-powered drones use convolutional neural networks (CNNs) to analyze images and classify vegetation. For example:

- ☆ **Crop vs. Weed Differentiation**: CNNs are trained on datasets containing images of various crops and weeds. The model learns to identify weeds based on unique features such as leaf shape, size, and color.
- ☆ **Species-Specific Identification**: Advanced AI models can identify specific weed species, enabling customized herbicide application.

2. Machine Learning for Pattern Analysis

ML algorithms analyze historical data to predict weed growth patterns and hotspots. These predictions help farmers implement proactive weed control measures. For instance:

- ☆ **Predictive Weed Mapping**: By analyzing past weed distribution maps, ML models can forecast weed emergence in specific areas.
- ☆ **Dynamic Thresholds**: ML can determine optimal weed density thresholds for intervention, balancing economic and environmental considerations.

3. Real-Time Decision Making

AI enables drones to process data in real-time, allowing immediate action. For example:

- ☆ **On-the-Fly Spraying**: Drones equipped with AI can detect weeds during flight and spray herbicides instantly, eliminating the need for pre-mapped spraying routes.

Applications in Modern Weed Management

The integration of drones with AI and ML has several practical applications in agriculture:

1. Precision Weed Control

- Drones equipped with AI-driven spraying systems can target weeds with centimeter-level accuracy, reducing herbicide wastage and preventing crop damage.

2. Cost-Effective Monitoring

- Regular drone surveys provide farmers with up-to-date weed maps, eliminating the need for manual scouting, which is labor-intensive and time-consuming.

3. Integrated Weed Management (IWM)

- AI-powered drones can be integrated into broader IWM strategies, combining mechanical, cultural, and chemical control methods for sustainable weed management.

4. Herbicide Resistance Management

- By identifying weed species and their distribution, AI systems help farmers rotate herbicides effectively, reducing the risk of resistance development.

Advantages of Drone-Based Weed Management

The use of drones integrated with AI and ML offers several advantages over traditional methods:

- **Environmental Sustainability**: Targeted herbicide application minimizes chemical runoff and soil contamination.
- **Labor Efficiency**: Automated weed detection and spraying reduce the need for manual labor.
- **Scalability**: Drones can cover large fields quickly, making them suitable for both small-scale and industrial agriculture.
- **Data-Driven Insights**: AI-generated weed maps and predictive models empower farmers to make informed decisions.

Challenges and Future Prospects

While the integration of drones with AI and ML holds immense potential, there are challenges to address:

- **High Initial Costs**: The cost of drones, sensors, and AI software can be prohibitive for small-scale farmers.

- **Data Processing Requirements**: High-resolution imagery requires significant computational power and storage.
- **Regulatory Constraints**: Drone usage is subject to local regulations, which may limit their deployment in certain regions.

Future advancements in AI algorithms, drone technology, and cost reduction strategies are expected to overcome these challenges, making these systems more accessible to farmers worldwide.

Case Study: AI-Driven Weed Management in Soybean Fields

A recent study conducted in the United States demonstrated the effectiveness of drones and AI in managing weeds in soybean fields. Researchers used drones equipped with multispectral cameras and AI algorithms to identify Palmer amaranth (Amaranthus palmeri), a highly invasive weed. The system achieved an accuracy rate of over 90 per cent in detecting the weed and significantly reduced herbicide usage by targeting only infested areas (Huang *et al.*, 2022).

I. Benefits of Drones and Remote Sensing in Weed Management

The integration of drones and remote sensing technologies into agricultural practices has revolutionized weed management. These advanced tools offer precision, efficiency, and cost-effectiveness, enabling farmers and agronomists to address weed infestations with greater accuracy and minimal environmental impact. Below, we explore the key benefits of drones and remote sensing in modern weed management.

1. Enhanced Weed Detection and Mapping

Drones equipped with high-resolution cameras and multispectral sensors provide detailed aerial imagery of agricultural fields. These images enable the identification and mapping of weed infestations with exceptional accuracy.

- **Precision Detection**: Multispectral and hyperspectral imaging can differentiate between crop plants and weeds based on their spectral signatures. For example, weeds often exhibit distinct reflectance patterns in the near-infrared (NIR) and red-edge bands.
- **Spatial Analysis**: Drones can generate georeferenced weed distribution maps, helping farmers identify hotspots of infestation and plan targeted interventions.
- **Temporal Monitoring**: Regular drone flights allow for time-series analysis, enabling farmers to monitor weed growth dynamics and evaluate the effectiveness of control measures.

In a study by Peña *et al.* (2013), drones were successfully used to detect broadleaf weeds in cereal crops, demonstrating their potential to enhance weed management strategies.

2. Precision Application of Herbicides

Drones equipped with sprayers can deliver herbicides with pinpoint accuracy, reducing chemical usage and minimizing environmental impact.

- **Site-Specific Application**: By integrating weed maps generated through remote sensing, drones can target only the infested areas, avoiding unnecessary herbicide application to healthy crops.
- **Variable Rate Technology (VRT)**: Drones can adjust the herbicide dosage based on weed density, ensuring optimal application rates.
- **Reduced Drift**: Low-altitude spraying by drones minimizes herbicide drift, protecting non-target plants and surrounding ecosystems.

This approach aligns with the principles of precision agriculture, promoting sustainable weed control practices.

3. Cost and Labor Efficiency

Traditional weed management methods, such as manual scouting and blanket herbicide application, are labor-intensive and time-consuming. Drones significantly reduce these challenges.

- **Time Savings**: A drone can survey large fields in a fraction of the time required for manual scouting.
- **Reduced Labor Costs**: By automating weed detection and herbicide application, drones lower the need for manual labor.
- **Economic Benefits**: Targeted herbicide application reduces chemical costs, while improved weed control enhances crop yields, resulting in higher profitability.

According to a report by Zhang and Kovacs (2012), the use of drones in agriculture can reduce operational costs by up to 30 per cent, making them a cost-effective solution for weed management.

4. Environmental Sustainability

Drones and remote sensing contribute to environmentally sustainable weed management practices by minimizing chemical inputs and reducing the ecological footprint of farming operations.

- **Minimized Herbicide Use**: Precision application reduces the total volume of herbicides applied, lowering the risk of soil and water contamination.
- **Biodiversity Conservation**: By targeting only weeds, drones help preserve beneficial flora and fauna in agricultural ecosystems.
- **Carbon Footprint Reduction**: The use of drones reduces the need for heavy machinery, decreasing fuel consumption and greenhouse gas emissions.

These benefits align with global efforts to promote sustainable agriculture and mitigate climate change.

5. Data-Driven Decision Making

Remote sensing technologies generate valuable data that can inform decision-making in weed management.

- ☆ **Predictive Analytics**: Historical data from drone surveys can be used to predict future weed outbreaks and plan proactive interventions.
- ☆ **Integration with Decision Support Systems (DSS)**: Weed maps and other data can be integrated into DSS platforms, enabling farmers to make informed decisions based on real-time insights.
- ☆ **Research Advancements**: The data collected by drones supports research on weed ecology, herbicide resistance, and the development of innovative control strategies.

For example, researchers have used drone-based remote sensing to study the spatial distribution of herbicide-resistant weeds, providing critical insights for resistance management (Huang *et al.*, 2018).

6. Accessibility and Scalability

The declining costs of drone technology and the availability of user-friendly software have made drones accessible to farmers of all scales.

- ☆ **Smallholder Farmers**: Affordable drones enable small-scale farmers to adopt precision weed management practices.
- ☆ **Scalability**: Drones can be deployed across various field sizes, from small plots to large commercial farms, making them a versatile tool for weed control.

J. Challenges in Using Drones and Remote Sensing

The integration of drones and remote sensing technologies in modern weed management has revolutionized precision agriculture. These tools provide real-time data, enabling farmers and researchers to identify weed infestations and make informed decisions. However, their adoption is not without challenges. This section explores the technical, operational, and regulatory hurdles in deploying drones and remote sensing for weed management.

1. Technical Challenges

- ☆ **Data Accuracy and Resolution:** While drones equipped with multispectral and hyperspectral sensors can capture detailed images, achieving high-resolution data for precise weed identification remains a challenge. Differentiating weeds from crops, especially in mixed vegetation or early growth stages, often requires advanced image processing algorithms and machine learning models, which may not always yield accurate results.

Example: Distinguishing broadleaf weeds from soybean crops using spectral indices like NDVI (Normalized Difference Vegetation Index) can be difficult due to overlapping spectral signatures (Zhang *et al.*, 2021).

- ☆ **Sensor Limitations:** The effectiveness of remote sensing depends on the quality of sensors. Low-cost drones often come with limited payload capacities, restricting the use of high-end sensors. Additionally, environmental factors such as cloud cover, lighting conditions, and atmospheric interference can degrade the quality of aerial imagery.
- ☆ **Data Processing and Analysis:** The vast amount of data collected by drones requires robust computational resources and specialized software for processing. For instance, stitching multiple images into an orthomosaic map or applying machine learning algorithms for weed detection can be time-intensive and computationally expensive.

2. Operational Challenges

- ☆ **Flight Time and Coverage Area:** Most drones have limited battery life, typically ranging from 20 to 40 minutes. This restricts their ability to cover large agricultural fields in a single flight. Frequent battery replacements or recharges can disrupt operations and increase costs.
- ☆ **Weather Dependency:** Drones are highly sensitive to weather conditions. High winds, rain, or extreme temperatures can hinder their performance, making it challenging to conduct consistent weed monitoring across seasons.
- ☆ **Skill Requirements:** Operating drones and interpreting remote sensing data require specialized training. Farmers and field technicians may lack the technical expertise needed to handle these advanced tools, necessitating additional investments in training programs.

3. Regulatory and Economic Challenges

- ☆ **Regulatory Restrictions:** Many countries have strict regulations governing drone usage, including restrictions on flight altitudes, no-fly zones, and licensing requirements. These regulations can limit the flexibility of drone operations in agricultural areas.

  Example: In the United States, the Federal Aviation Administration (FAA) mandates that drones must be operated within the visual line of sight of the pilot, which can be a constraint for large-scale farms.
- ☆ **High Initial Investment:** The cost of acquiring drones, sensors, and data processing software can be prohibitive for small and medium-scale

farmers. While the long-term benefits of precision weed management are significant, the upfront costs can deter widespread adoption.

- ✰ **Data Privacy and Security:** The use of drones raises concerns about data privacy and security. Unauthorized access to aerial imagery could potentially expose sensitive information about farm operations, leading to reluctance among farmers to adopt these technologies.

4. Environmental and Ecological Concerns

- ✰ **Disturbance to Wildlife:** The noise generated by drones can disturb wildlife in agricultural ecosystems. This is particularly relevant in areas where farms coexist with natural habitats.
- ✰ **Energy Consumption:** The frequent charging of drone batteries and the computational power required for data processing contribute to the overall carbon footprint of these technologies, raising questions about their sustainability.

K. Future Innovations and Trends for Drones and Remote Sensing

The integration of drones and remote sensing technologies in modern weed management is rapidly evolving, driven by advancements in sensor capabilities, artificial intelligence (AI), and data analytics. These innovations are revolutionizing how weeds are detected, monitored, and managed, offering precision, efficiency, and sustainability.

1. Enhanced Sensor Technologies

- ✰ **Multispectral and Hyperspectral Imaging**: Future drones will increasingly utilize advanced multispectral and hyperspectral sensors to detect subtle differences in plant reflectance. These sensors can differentiate between crop species and weed species based on their unique spectral signatures, enabling precise weed identification.
- ✰ **Thermal Imaging**: Thermal sensors are being developed to detect temperature variations caused by water stress or metabolic activity in weeds, providing another layer of differentiation from crops.
- ✰ **LiDAR (Light Detection and Ranging)**: LiDAR technology is emerging as a tool for creating 3D maps of fields, which can be used to identify weed-infested areas based on vegetation height and density.

2. AI-Powered Weed Detection and Analysis

- ✰ **Machine Learning Algorithms**: AI-driven algorithms are becoming more sophisticated in processing drone-acquired imagery. These algorithms can classify weeds with high accuracy, even in complex field conditions, by learning from large datasets of annotated images.

- **Real-Time Processing**: Onboard AI systems are being developed to process data in real time, enabling immediate weed detection and targeted action during drone flights.

3. Precision Weed Control

- **Variable Rate Herbicide Application**: Drones equipped with precision sprayers are being designed to apply herbicides only to weed-infested areas, reducing chemical usage and minimizing environmental impact.
- **Autonomous Swarm Drones**: The concept of multiple drones working collaboratively is gaining traction. Swarm drones can cover large fields efficiently, with each drone assigned specific tasks, such as weed mapping or herbicide application.

4. Integration with IoT and Big Data

- **Internet of Things (IoT)**: Drones are increasingly being integrated with IoT devices, such as soil moisture sensors and weather stations, to provide a holistic view of field conditions. This integration allows for more informed weed management decisions.
- **Big Data Analytics**: The vast amount of data collected by drones can be analyzed using big data tools to identify weed infestation trends, predict outbreaks, and optimize management strategies.

5. Regulatory and Ethical Considerations

- **Data Privacy**: As drones collect detailed field data, ensuring data privacy and security will become a critical focus.
- **Regulatory Frameworks**: Governments are likely to introduce stricter regulations for drone usage in agriculture, emphasizing safety and environmental compliance.

6. Future Challenges and Opportunities

- **Battery Life and Payload Capacity**: Innovations in drone battery technology and lightweight materials are expected to address current limitations in flight time and payload capacity.
- **Cost Reduction**: As technology matures, the cost of drones and sensors is expected to decrease, making these tools accessible to small and medium-scale farmers.

Chapter 7

Sensor-Based Weed Control Systems

Sensor-based weed control systems represent a cutting-edge approach to precision agriculture, leveraging advanced technologies to detect and manage weeds with high efficiency. These systems utilize sensors to differentiate between crops and weeds based on various parameters such as spectral reflectance, shape, or thermal properties. By enabling targeted herbicide application, they significantly reduce chemical use, operational costs, and environmental impact.

Key Components of Sensor-Based Systems

1. Sensors

- ☆ **Optical Sensors**: Detect differences in light reflectance between crops and weeds, often using multispectral or hyperspectral imaging.
- ☆ **Thermal Sensors**: Identify temperature variations caused by differences in plant transpiration rates.
- ☆ **LIDAR (Light Detection and Ranging)**: Measures plant height and structure to distinguish weeds from crops.

2. Data Processing Units

Analyze sensor data in real-time using machine learning algorithms to classify vegetation accurately.

3. Actuators

Control mechanisms such as sprayers or mechanical weeders to act precisely on identified weed patches.

Applications

- **Green-on-Brown Systems**: Detect weeds on bare soil, commonly used in pre-emergence weed control.
- **Green-on-Green Systems**: Differentiate weeds from crops in dense vegetation, a more complex but highly effective approach.

Recent Advances

- Integration of **AI and machine learning** has enhanced weed identification accuracy.
- Development of **autonomous robotic weeders** equipped with sensor-based systems for continuous field operations (Slaughter *et al.*, 2008).

Challenges

- High initial costs and the need for skilled operators.
- Limitations in distinguishing weeds from crops with similar spectral or structural characteristics.

Introduction to Sensor-Based Weed Control Systems

Sensor-based weed control systems represent a transformative approach in modern agriculture, leveraging advanced technologies to enhance weed management efficiency. These systems utilize sensors to detect, identify, and target weeds with precision, reducing herbicide usage, minimizing environmental impact, and improving crop productivity.

Principles of Sensor-Based Weed Control

Sensor-based weed control relies on the integration of sensing technologies with automated machinery. The key principle involves detecting differences between weeds and crops based on specific parameters such as spectral reflectance, shape, size, or thermal properties. Sensors collect real-time data, which is processed by algorithms to distinguish weeds from crops. Once identified, targeted control measures, such as spot spraying or mechanical removal, are applied.

Sensors Used in Weed Control

Several types of sensors are employed in weed control systems, each with unique capabilities:

- **Optical Sensors**: These sensors detect differences in light reflectance between weeds and crops. For example, near-infrared (NIR) sensors can differentiate plant species based on their spectral signatures. Optical sensors are widely used in systems like the WeedSeeker® and Weed-It® for precision herbicide application.

- **Multispectral and Hyperspectral Sensors**: These advanced sensors capture data across multiple wavelengths, enabling detailed weed-crop differentiation. They are particularly effective in identifying weed species in complex crop environments.
- **Thermal Sensors**: By detecting temperature variations, thermal sensors can identify weeds based on their transpiration rates or canopy temperatures.
- **Lidar (Light Detection and Ranging)**: Lidar sensors use laser pulses to create 3D maps of the field, helping to distinguish weeds from crops based on their structural characteristics.
- **Machine Vision Systems**: These systems use cameras and image processing algorithms to identify weeds based on shape, size, and texture.

Applications and Benefits

Sensor-based weed control systems are primarily used in precision agriculture, where site-specific weed management is critical. Key benefits include:

- **Reduced Herbicide Usage**: By targeting only the weeds, these systems significantly lower herbicide consumption, reducing costs and environmental risks.
- **Improved Crop Health**: Minimizing herbicide exposure to crops enhances their growth and yield potential.
- **Environmental Sustainability**: Precision application reduces chemical runoff, protecting soil and water quality.
- **Labor Efficiency**: Automated systems reduce the need for manual labor, particularly in large-scale farming operations.

Challenges and Future Directions

Despite their advantages, sensor-based weed control systems face challenges such as high initial costs, complexity in distinguishing weeds in dense crop canopies, and the need for robust algorithms to handle diverse field conditions. Future advancements in artificial intelligence (AI) and machine learning (ML) are expected to enhance the accuracy and adaptability of these systems.

Example Systems

- **Blue River Technology's See and Spray**: This system uses AI-powered cameras to identify and spray weeds selectively, reducing herbicide use by up to 90 per cent.
- **WeedSeeker®**: An optical sensor-based system that detects green plants and applies herbicide only where needed.

A. Types of Sensors Used in Weed Control

The integration of advanced sensor technologies in weed control has revolutionized modern agriculture, enabling precise, efficient, and environmentally sustainable weed management practices. Sensors play a critical role in detecting, identifying, and monitoring weeds in real-time, facilitating site-specific herbicide application and reducing chemical usage. Below are the primary types of sensors used in weed control, categorized based on their functionality and application.

1. Optical Sensors

Optical sensors are among the most widely used technologies in weed detection. These sensors operate by analyzing the interaction of light with plant surfaces, allowing for the differentiation between crops and weeds.

- ✰ **Active Optical Sensors**: These sensors emit their own light source, typically in the visible or near-infrared (NIR) spectrum, and measure the reflected light to identify plant characteristics. For example, the GreenSeeker and WeedSeeker systems use active optical sensors to detect weeds and apply herbicides only where necessary.
- ✰ **Passive Optical Sensors**: These rely on ambient sunlight to capture spectral data. Multispectral and hyperspectral cameras are examples of passive optical sensors that can distinguish weeds from crops based on their unique spectral signatures.

Scientific Principle: Optical sensors leverage the differences in reflectance and absorption of light by chlorophyll and other plant pigments. For instance, weeds often exhibit distinct spectral patterns compared to crops, enabling accurate identification (Mulla, 2013).

2. Thermal Sensors

Thermal sensors detect temperature variations between plants and their surroundings. Weeds and crops often exhibit different thermal properties due to variations in water content, leaf structure, and transpiration rates.

- ✰ **Infrared Cameras**: These sensors capture thermal images to identify weeds based on their heat signatures.
- ✰ **Applications**: Thermal sensors are particularly useful in low-light conditions or at night, providing continuous monitoring capabilities.

Scientific Principle: Thermal sensors exploit the differences in emissivity and thermal conductivity between plant species. For example, weeds with higher transpiration rates may appear cooler in thermal imagery.

3. Proximity Sensors

Proximity sensors are used to detect the presence of plants within a specific range. These sensors are commonly employed in mechanical weed control systems.

- ☆ **Ultrasonic Sensors**: These sensors emit sound waves and measure the time taken for the waves to return after hitting a plant surface. They are effective in distinguishing between tall weeds and shorter crops.
- ☆ **LIDAR (Light Detection and Ranging)**: LIDAR sensors use laser pulses to create 3D maps of the field, identifying weeds based on their height, shape, and spatial distribution.

Scientific Principle: Proximity sensors rely on the physical differences between crops and weeds, such as height and density, to enable targeted interventions.

4. Electrochemical Sensors

Electrochemical sensors detect specific chemical compounds released by plants. These sensors are used to identify weeds based on their unique biochemical profiles.

- ☆ **Applications**: Electrochemical sensors can measure the presence of allelopathic compounds or other metabolites that are specific to certain weed species.

Scientific Principle: Weeds often produce distinct secondary metabolites that can be detected electrochemically, aiding in species-specific weed management.

5. Machine Vision Sensors

Machine vision systems, often integrated with artificial intelligence (AI), use cameras and advanced algorithms to identify and classify weeds.

- ☆ **RGB Cameras**: These standard cameras capture high-resolution images for weed identification based on color and texture.
- ☆ **AI Integration**: Deep learning models process the captured images to distinguish weeds from crops with high accuracy.

Scientific Principle: Machine vision systems rely on pattern recognition and image processing techniques to analyze plant morphology and differentiate between species.

6. Soil Sensors

Soil sensors monitor soil properties, such as moisture, temperature, and nutrient levels, to predict weed emergence and growth patterns.

- ☆ **Applications**: These sensors are used in predictive weed management, allowing farmers to take preventive measures before weed infestations occur.

Scientific Principle: Soil conditions significantly influence weed germination and growth. By monitoring these parameters, soil sensors help in developing proactive weed control strategies.

B. Applications of Sensor-Based Systems in Weed Control

The integration of sensor-based systems into weed management has revolutionized modern agricultural practices by enabling precise, efficient, and sustainable weed control. These systems leverage advanced technologies such as optical sensors, multispectral imaging, and machine learning algorithms to detect, identify, and manage weeds in real-time. This section explores the applications of sensor-based systems in weed control, highlighting their functionality, benefits, and examples of their use in modern agriculture.

1. Principles of Sensor-Based Weed Detection

Sensor-based systems rely on the ability to differentiate between crops and weeds based on their unique physical, spectral, or chemical properties. These systems utilize various types of sensors, including:

- ✰ **Optical Sensors**: Detect differences in light reflectance between crops and weeds. For example, weeds often reflect light differently in the near-infrared (NIR) spectrum compared to crops.
- ✰ **Multispectral and Hyperspectral Sensors**: Capture data across multiple wavelengths to identify specific weed species based on their spectral signatures.
- ✰ **Thermal Sensors**: Measure temperature differences between plants, as weeds may exhibit distinct thermal properties due to differences in water content or leaf structure.
- ✰ **LiDAR (Light Detection and Ranging)**: Generates 3D maps of the field to identify weed growth patterns based on plant height and structure.

These sensors are often mounted on drones, tractors, or autonomous robots to scan fields and collect data in real-time.

2. Site-Specific Weed Management (SSWM)

Sensor-based systems enable **Site-Specific Weed Management (SSWM)**, which focuses on applying weed control measures only where they are needed. This approach minimizes herbicide use, reduces costs, and lowers environmental impact. Key applications include:

- ✰ **Weed Mapping**: Sensors collect data to create detailed weed distribution maps, allowing farmers to target specific areas with high weed density.
- ✰ **Variable Rate Herbicide Application**: Using weed maps, sprayers equipped with sensors can adjust the herbicide dosage based on weed density and type, ensuring precise application.

- **Mechanical Weed Control**: Autonomous robots equipped with sensors can identify and mechanically remove weeds without damaging crops.

For instance, research has shown that using sensor-based systems for variable rate herbicide application can reduce herbicide usage by up to 70 per cent without compromising weed control efficacy (Slaughter *et al.*, 2008).

3. Real-Time Weed Identification

Modern sensor-based systems incorporate machine learning algorithms to classify weeds in real-time. These systems analyze data from sensors to distinguish weeds from crops based on features such as:

- Leaf shape
- Color
- Growth patterns
- Spectral reflectance

For example, the "See and Spray" technology developed by Blue River Technology uses high-resolution cameras and machine learning to identify and spray individual weeds with herbicides, significantly reducing chemical usage.

4. Integration with Autonomous Systems

Sensor-based weed control systems are increasingly integrated with autonomous platforms such as drones and robotic weeders. These platforms enhance efficiency and scalability in large-scale farming operations. Applications include:

- **Drone-Based Weed Monitoring**: Drones equipped with multispectral sensors can survey large fields quickly, providing high-resolution weed maps.
- **Autonomous Weed Removal**: Robots such as the "EcoRobotix" use sensors to detect weeds and apply precise doses of herbicide or mechanically remove them.

5. Benefits of Sensor-Based Weed Control

The adoption of sensor-based systems offers several advantages over traditional weed control methods:

- **Precision**: Accurate weed detection and targeted control reduce the risk of crop damage.
- **Cost-Effectiveness**: Lower herbicide usage translates to reduced input costs.
- **Environmental Sustainability**: Minimizing chemical application reduces the risk of soil and water contamination.

- ☆ **Labor Efficiency**: Automation reduces the need for manual labor in weed management.

6. Challenges and Future Prospects

While sensor-based systems hold immense potential, challenges such as high initial costs, the need for technical expertise, and limitations in detecting weeds under dense crop canopies remain. However, ongoing advancements in sensor technology, artificial intelligence, and robotics are expected to address these challenges, making sensor-based weed control more accessible and effective.

C. Real-Time Weed Detection and Monitoring

The integration of real-time weed detection and monitoring technologies into modern agricultural practices has revolutionized weed management. These systems leverage advancements in sensors, machine vision, artificial intelligence (AI), and robotics to identify and monitor weed populations with unprecedented accuracy and efficiency. By enabling site-specific weed control, these technologies reduce herbicide use, minimize environmental impacts, and enhance crop productivity.

Technologies Enabling Real-Time Weed Detection

1. Machine Vision Systems

Machine vision systems use cameras and image processing algorithms to distinguish weeds from crops based on morphological, spectral, and textural differences. These systems rely on high-resolution cameras mounted on tractors, drones, or robotic platforms to capture field images.

- ☆ **Multispectral and Hyperspectral Imaging**: These imaging systems capture data across multiple wavelengths, allowing for the differentiation of plant species based on their unique spectral signatures. For instance, weeds often exhibit distinct reflectance patterns in the near-infrared (NIR) spectrum compared to crops.
- ☆ **Deep Learning Algorithms**: Convolutional Neural Networks (CNNs) are widely used to classify plants in real-time. These algorithms are trained on large datasets of annotated images, enabling them to identify specific weed species with high accuracy.

2. Proximal Sensors

Proximal sensors, such as LiDAR (Light Detection and Ranging) and ultrasonic sensors, are employed to detect the physical structure and height of plants. These sensors are particularly useful in distinguishing weeds from crops based on their growth patterns and canopy architecture.

3. Robotics and Automation

Autonomous robots equipped with weed detection systems are increasingly being deployed in precision agriculture. These robots can navigate fields, identify weeds, and perform targeted control measures such as mechanical weeding or spot spraying. For example, the "See and Spray" technology developed by Blue River Technology combines machine vision with robotics to apply herbicides only to detected weeds, significantly reducing chemical usage.

Applications of Real-Time Weed Monitoring

1. Site-Specific Weed Management (SSWM)

Real-time weed detection enables site-specific weed management by mapping weed infestations and applying control measures only where necessary. This approach minimizes over-application of herbicides, reduces costs, and protects non-target organisms.

2. Weed Population Mapping

Drones equipped with imaging systems can survey large fields and generate detailed weed distribution maps. These maps provide insights into weed density, species composition, and spatial patterns, aiding in the development of targeted management strategies.

3. Early Detection of Herbicide Resistance

Real-time monitoring systems can identify changes in weed populations that may indicate the development of herbicide resistance. For example, a sudden increase in the density of a particular weed species despite herbicide application can signal resistance, prompting timely intervention.

Challenges in Real-Time Weed Detection

Despite its potential, real-time weed detection faces several challenges:

- **Complexity of Weed-Crop Differentiation**: Variability in plant morphology, overlapping canopies, and similar spectral properties between weeds and crops can complicate detection.
- **Environmental Factors**: Lighting conditions, soil background, and weather can affect the accuracy of imaging systems.
- **Cost and Scalability**: High initial costs of advanced sensors and robotics may limit adoption, particularly for small-scale farmers.

Future Directions

Advancements in AI, edge computing, and sensor technologies are expected to address current limitations and enhance the performance of real-time weed detection systems. For instance, integrating Internet of Things (IoT) devices with weed monitoring platforms can enable real-time data sharing and decision-making across farm operations.

Case Study Example

A study conducted by Milioto *et al.* (2021) demonstrated the effectiveness of deep learning-based weed detection in sugar beet fields. The researchers used a CNN model trained on field images to achieve over 90 per cent accuracy in distinguishing weeds from crops, highlighting the potential of AI-driven approaches in precision agriculture.

D. Integration with Spraying Systems for Sensor-Based Weed Control Systems

The integration of sensor-based weed control systems with advanced spraying technologies has revolutionized modern weed management practices. This approach combines precision agriculture techniques with automated spraying systems to optimize herbicide application, reduce chemical usage, and improve environmental sustainability. By leveraging real-time data and advanced algorithms, these systems enable site-specific weed control, ensuring that herbicides are applied only where needed. This section explores the principles, components, and operational aspects of integrating sensor-based weed control systems with spraying technologies.

Principles of Sensor-Based Weed Control Systems

Sensor-based weed control systems rely on advanced sensing technologies to detect and differentiate weeds from crops. These systems use optical, thermal, or multispectral sensors to gather real-time data on plant morphology, spectral reflectance, and other distinguishing characteristics. The key principles include:

- ☆ **Weed-Crop Differentiation**: Sensors identify weeds based on their unique spectral signatures, leaf shapes, or growth patterns.
- ☆ **Real-Time Data Processing**: Algorithms process sensor data in real time to determine the exact location and density of weeds.
- ☆ **Targeted Application**: Once weeds are identified, the system triggers the spraying mechanism to apply herbicides precisely to the target area.

For example, hyperspectral imaging sensors can distinguish between crop and weed species based on their reflectance in specific wavelengths, enabling accurate weed mapping and targeted control.

Components of an Integrated System

The integration of sensor-based weed detection with spraying systems involves several interconnected components:

1. Sensors

- ✰ Optical sensors (*e.g.*, RGB cameras, hyperspectral sensors) detect weed presence and characteristics.
- ✰ LiDAR (Light Detection and Ranging) sensors provide spatial data for weed localization.
- ✰ Thermal sensors detect temperature differences between weeds and crops.

2. Control Unit

- ✰ The central processing unit (CPU) analyzes sensor data using machine learning algorithms or pre-programmed weed identification models.
- ✰ Decision-making software determines the appropriate herbicide dosage and spraying pattern.

3. Spraying System

- ✰ Equipped with variable-rate nozzles, the spraying system adjusts herbicide flow rates based on weed density and type.
- ✰ Boom sprayers with sectional or individual nozzle control ensure precise application.

4. GPS and Mapping Tools

- ✰ GPS technology provides geospatial data for accurate weed location mapping.
- ✰ Weed maps generated by the system guide the spraying operation.

5. Connectivity and Data Storage

- ✰ Cloud-based platforms store weed detection data for future analysis and decision-making.
- ✰ Wireless communication enables real-time system updates and remote monitoring.

Operational Workflow

The integration of sensor-based weed control with spraying systems follows a systematic workflow:

1. Weed Detection

- ✰ Sensors mounted on tractors, drones, or autonomous vehicles scan the field to detect weeds.
- ✰ Data is transmitted to the control unit for analysis.

2. **Data Processing**

- ☆ Algorithms classify plants as weeds or crops based on predefined criteria.
- ☆ Weed density and spatial distribution are mapped.

3. **Spraying Activation**

- ☆ The control unit sends commands to the spraying system to activate specific nozzles.
- ☆ Variable-rate technology adjusts the herbicide dosage according to weed density.

4. **Herbicide Application**

- ☆ Herbicides are applied with high precision, minimizing waste and off-target effects.
- ☆ The system records application data for compliance and analysis.

Advantages of Integration

The integration of sensor-based weed control systems with spraying technologies offers several benefits:

- ☆ **Reduced Herbicide Usage**: Targeted application minimizes chemical use, reducing costs and environmental impact.
- ☆ **Improved Weed Control Efficiency**: Real-time detection and spraying ensure effective weed management, even in dense infestations.
- ☆ **Environmental Sustainability**: Reduced chemical runoff and drift protect soil and water resources.
- ☆ **Labor Savings**: Automation reduces the need for manual weed scouting and spraying.
- ☆ **Data-Driven Decision Making**: Weed maps and application records enable informed management decisions.

Challenges and Considerations

Despite its advantages, the integration of sensor-based weed control systems with spraying technologies presents certain challenges:

- ☆ **High Initial Costs**: The cost of sensors, control units, and advanced spraying systems can be prohibitive for small-scale farmers.
- ☆ **Complexity of Weed Identification**: Variability in weed species, growth stages, and environmental conditions can affect detection accuracy.
- ☆ **Maintenance and Calibration**: Regular maintenance and calibration of sensors and spraying equipment are essential for optimal performance.

- ☆ **Data Management**: Large volumes of data generated by the system require robust storage and processing capabilities.

Examples of Commercial Systems

Several commercial systems demonstrate the successful integration of sensor-based weed control with spraying technologies:

John Deere See and Spray™

- ☆ Combines machine vision and artificial intelligence to detect and spray weeds selectively.
- ☆ Reduces herbicide use by up to 77 per cent compared to traditional broadcast spraying.

Blue River Technology

- ☆ Uses computer vision and machine learning to identify weeds and apply herbicides precisely.
- ☆ Integrated with advanced sprayers for real-time operation.

WEEDit

- ☆ Employs near-infrared (NIR) sensors to detect chlorophyll in weeds.
- ☆ Activates nozzles only when weeds are present, ensuring precise application.

Future Prospects

The integration of sensor-based weed control systems with spraying technologies is expected to evolve further with advancements in artificial intelligence, robotics, and IoT (Internet of Things). Emerging trends include autonomous weed control robots, drone-based spraying systems, and the use of big data analytics for predictive weed management.

E. Role of Sensor-Based Weed Control Systems in Autonomous Weed Control Machines

The integration of sensor-based weed control systems into autonomous weed control machines represents a significant advancement in modern agricultural practices. These systems leverage cutting-edge technologies to enhance weed management efficiency, reduce herbicide usage, and promote sustainable farming. By enabling precise weed detection and targeted control, sensor-based systems are transforming the landscape of weed management.

1. Overview of Sensor-Based Weed Control Systems

Sensor-based weed control systems utilize advanced sensors to detect, identify, and differentiate weeds from crops. These systems are typically integrated into autonomous machines, such as robotic weeders or drones, to enable real-time decision-making and action.

Key Components of Sensor-Based Systems

- **Sensors**: Devices such as cameras, LiDAR, hyperspectral sensors, and multispectral sensors are used to capture data on plant characteristics.
- **Processing Units**: Algorithms and machine learning models process the sensor data to classify plants as weeds or crops.
- **Actuators**: Mechanisms that execute weed control actions, such as spraying herbicides, mechanical weeding, or laser-based weed removal.

Advantages

- High precision in distinguishing weeds from crops.
- Reduction in herbicide application, minimizing environmental impact.
- Compatibility with various crop types and field conditions.

2. Types of Sensors Used in Weed Control

a) Optical Sensors

Optical sensors, such as RGB cameras and hyperspectral imaging systems, are widely used in weed detection. These sensors capture visual data, which is analyzed to identify plant features like shape, size, and color.

- **RGB Cameras**: Provide high-resolution images for basic weed-crop differentiation.
- **Hyperspectral Sensors**: Capture data across a wide spectrum of wavelengths, enabling the identification of subtle differences in plant reflectance.

b) LiDAR Sensors

LiDAR (Light Detection and Ranging) sensors use laser pulses to measure distances and create 3D maps of the field. These maps help in identifying the spatial distribution of weeds and crops.

c) Thermal Sensors

Thermal imaging sensors detect temperature variations between weeds and crops. This can be particularly useful in distinguishing plants with different transpiration rates or water content.

d) Multispectral Sensors

Multispectral sensors capture data in specific wavelength bands, such as red, green, blue, and near-infrared. These sensors are effective in identifying plant health and stress levels, aiding in weed detection.

3. Machine Learning and Artificial Intelligence in Weed Detection

Sensor-based systems rely heavily on machine learning (ML) and artificial intelligence (AI) to process and interpret data. These technologies enable the development of robust models for weed identification and classification.

a) Image Processing Algorithms

- Algorithms analyze images captured by sensors to extract features such as leaf shape, texture, and color.
- Techniques like edge detection and segmentation are used to isolate weeds from crops.

b) Deep Learning Models

- Convolutional Neural Networks (CNNs) are commonly employed for image recognition tasks.
- These models are trained on large datasets of weed and crop images to achieve high accuracy in classification.

c) Real-Time Decision Making

- AI systems enable autonomous machines to make real-time decisions on weed control actions.
- For example, a robotic weeder can decide whether to spray herbicide, mechanically remove a weed, or leave the plant untouched.

Example: A study by Milioto *et al.* (2018) demonstrated the use of deep learning for weed detection in sugar beet fields, achieving over 90 per cent accuracy in distinguishing weeds from crops.

4. Applications in Autonomous Weed Control Machines

Autonomous weed control machines equipped with sensor-based systems are revolutionizing weed management. These machines operate independently, reducing the need for manual labor and improving operational efficiency.

a) Robotic Weeders

- Robotic weeders use sensors to navigate fields and target weeds with precision.
- Example: The "Naïo Dino" robot employs GPS and camera-based sensors to autonomously weed vegetable crops.

b) Drones

- Drones equipped with multispectral or hyperspectral sensors can survey large fields to map weed infestations.
- These maps guide targeted weed control measures, such as spot spraying.

c) Laser-Based Weed Control

- Some machines use lasers to destroy weeds without harming crops.
- Sensors guide the laser to the exact location of the weed, ensuring precision.

5. Benefits of Sensor-Based Weed Control Systems

The adoption of sensor-based systems in autonomous weed control machines offers several benefits:

- **Cost Efficiency**: Reduced herbicide usage and labor costs.
- **Environmental Sustainability**: Minimized chemical runoff and soil degradation.
- **Improved Crop Yields**: Enhanced weed control leads to healthier crops and higher yields.
- **Scalability**: Suitable for both small-scale and large-scale farming operations.

Economic Impact

A study by Slaughter *et al.* (2008) highlighted that sensor-based precision weed control can reduce herbicide costs by up to 70 per cent, making it a cost-effective solution for farmers.

6. Challenges and Future Directions

Despite their advantages, sensor-based weed control systems face certain challenges:

a) High Initial Costs

- The cost of sensors and autonomous machines can be prohibitive for small-scale farmers.

b) Data Processing Requirements

- Advanced computational resources are needed to process large volumes of sensor data in real-time.

c) Field Variability

- Variations in soil, lighting, and crop types can affect sensor performance.

Future Trends

- Integration of Internet of Things (IoT) for real-time data sharing.
- Development of low-cost sensors to make the technology accessible to smallholders.
- Enhanced machine learning models for better weed-crop differentiation.

F. Benefits of Sensor-Based Weed Control

Sensor-based weed control is an innovative and precise approach to managing weeds in modern agriculture. By integrating advanced sensor technologies with automated systems, this method offers numerous advantages over traditional weed management techniques. Below are the key benefits of sensor-based weed control:

1. Enhanced Precision in Weed Detection

- ☆ Sensor-based systems use advanced technologies such as optical sensors, multispectral cameras, and LiDAR to identify weeds with high accuracy. These sensors can differentiate between crops and weeds based on shape, size, color, and spectral reflectance.
- ☆ Precision weed detection minimizes the risk of crop damage during herbicide application or mechanical weeding, ensuring that only the targeted weeds are affected.

2. Reduced Herbicide Usage

- ☆ By targeting weeds selectively, sensor-based systems significantly reduce the volume of herbicides required. For example, spot-spraying systems equipped with sensors apply herbicides only where weeds are detected, avoiding blanket applications across the entire field.
- ☆ Studies have shown that sensor-based weed control can reduce herbicide usage by up to 70 per cent compared to conventional methods (Slaughter *et al.*, 2008). This not only lowers input costs but also minimizes environmental contamination.

3. Cost Efficiency

- ☆ Reduced herbicide usage directly translates into lower operational costs for farmers. Additionally, the precise application of herbicides reduces the need for re-treatment, saving time and labor expenses.
- ☆ Over time, the investment in sensor-based systems can be offset by the cumulative savings in chemical inputs and labor.

4. Environmental Sustainability

- ☆ Sensor-based weed control aligns with the principles of sustainable agriculture by reducing the environmental footprint of weed management practices. Lower herbicide usage decreases the risk of soil and water contamination, protecting ecosystems and biodiversity.
- ☆ By minimizing chemical inputs, this method also helps in mitigating herbicide resistance in weed populations, a growing concern in modern agriculture.

5. Improved Crop Yields

- Weeds compete with crops for essential resources such as sunlight, water, and nutrients. Sensor-based systems enable timely and effective weed control, reducing competition and ensuring optimal growing conditions for crops.
- Enhanced precision in weed management can lead to healthier crops and higher yields, contributing to improved farm profitability.

6. Real-Time Data Collection and Analysis

- Many sensor-based systems are integrated with GPS and data analytics tools, enabling real-time monitoring of weed populations. This data can be used to create weed density maps, track weed growth patterns, and assess the effectiveness of control measures.
- The ability to analyze data over time helps farmers make informed decisions and refine their weed management strategies.

7. Applicability Across Diverse Farming Systems

- Sensor-based weed control systems are adaptable to various farming practices, including conventional, organic, and conservation agriculture. For instance, optical sensors can be used in organic systems to guide mechanical weeding tools, while in conventional systems, they can optimize herbicide application.
- These systems are also compatible with different crop types and field conditions, making them versatile tools for modern agriculture.

8. Labor Efficiency

- Traditional weed control methods, such as manual weeding, are labor-intensive and time-consuming. Sensor-based systems automate the weed detection and control process, reducing the reliance on manual labor.
- This is particularly beneficial in regions facing labor shortages or high labor costs.

9. Integration with Smart Farming Technologies

- Sensor-based weed control can be seamlessly integrated into broader precision agriculture frameworks. For example, data from weed sensors can be combined with soil moisture sensors, weather data, and crop health monitoring systems to optimize overall farm management.
- The integration of these technologies enhances the efficiency and effectiveness of agricultural operations.

10. Long-Term Economic and Environmental Benefits

- By promoting sustainable practices and reducing input costs, sensor-based weed control contributes to the long-term economic viability of farms.
- The environmental benefits, such as reduced chemical runoff and preservation of soil health, ensure that agricultural lands remain productive for future generations.

G. Challenges in Sensor Implementation

The integration of sensor technology in modern weed management has revolutionized precision agriculture, offering real-time data collection and improved decision-making. However, the implementation of sensors in agricultural systems is not without challenges. These challenges can be broadly categorized into technical, environmental, economic, and operational aspects.

1. Technical Challenges

- **Accuracy and Calibration**: Sensors require precise calibration to ensure accurate data collection. Variability in sensor performance due to environmental conditions, such as temperature fluctuations or soil moisture, can lead to errors in weed detection.
- **Data Integration**: Modern weed management often involves multiple sensors (*e.g.*, optical, thermal, and hyperspectral). Integrating data from these diverse sources into a unified decision-making platform is complex and requires advanced algorithms and software.
- **Power Supply**: Many sensors, especially those deployed in remote fields, rely on batteries or solar power. Ensuring a consistent power supply for uninterrupted operation remains a challenge.

2. Environmental Challenges

- **Field Variability**: Heterogeneous field conditions, such as varying soil types, crop residues, and weed densities, can affect sensor performance. For example, optical sensors may struggle to differentiate between weeds and crops in dense vegetation.
- **Weather Conditions**: Adverse weather, such as heavy rain, fog, or dust, can interfere with sensor readings. For instance, LiDAR sensors may experience reduced accuracy in foggy conditions due to light scattering (Slaughter *et al.*, 2008).
- **Biological Factors**: The dynamic nature of weed growth stages and morphology can complicate detection. Sensors must adapt to these changes to maintain effectiveness.

3. Economic Challenges

- ✰ **High Initial Costs**: Advanced sensors, such as hyperspectral imaging systems, are expensive to acquire and maintain. This can be a barrier for small and medium-scale farmers.
- ✰ **Cost-Benefit Analysis**: Farmers must weigh the benefits of sensor-based weed management against traditional methods. In some cases, the return on investment may not justify the costs.

4. Operational Challenges

- ✰ **User Training**: Effective use of sensor technology requires specialized knowledge. Farmers and operators must be trained to interpret sensor data and integrate it into weed management strategies.
- ✰ **Maintenance and Durability**: Sensors deployed in agricultural fields are exposed to harsh conditions, including dust, water, and mechanical impacts. Ensuring durability and ease of maintenance is critical for long-term use.

5. Data Privacy and Security

- ✰ **Data Ownership**: The use of sensors often involves the collection of large amounts of data. Questions about data ownership and privacy can arise, particularly when third-party service providers are involved.
- ✰ **Cybersecurity Risks**: As sensors become increasingly connected to cloud-based platforms, they are vulnerable to cyberattacks, which could compromise sensitive agricultural data.

In summary, while sensor technology has immense potential to enhance weed management, addressing these challenges is crucial for its successful implementation. Continued research and development, along with farmer education, will play a key role in overcoming these barriers.

H. Advancements in Sensor Technology

Sensor technology has revolutionized modern weed management by enabling precise, data-driven decisions in agricultural practices. The integration of advanced sensors with agricultural machinery and digital platforms has significantly enhanced the efficiency of weed detection, monitoring, and control. This section explores the latest advancements in sensor technology and their applications in weed management.

Types of Sensors in Weed Management

Modern weed management employs a variety of sensors, each designed to address specific challenges in detecting and controlling weeds. Key types include:

- **Optical Sensors**: These sensors use light reflectance and absorption properties to distinguish weeds from crops. Multispectral and hyperspectral sensors, for instance, capture data across multiple wavelengths, enabling the identification of weed species based on their unique spectral signatures.
- **Thermal Sensors**: Thermal imaging cameras detect temperature differences between weeds and crops, which can be useful in identifying weeds under varying environmental conditions.
- **LiDAR (Light Detection and Ranging)**: LiDAR sensors create high-resolution 3D maps of fields, allowing for precise weed mapping and monitoring.
- **Electromagnetic Sensors**: These sensors measure soil conductivity and other electromagnetic properties to detect weed-infested areas indirectly.

Applications of Sensor Technology in Weed Management

The integration of sensors into agricultural systems has enabled several innovative applications, including:

1. Weed Detection and Mapping

Advanced sensors can identify weeds at an early growth stage, allowing for timely interventions. For example, hyperspectral imaging can differentiate between crop and weed species based on their spectral reflectance patterns. This data is often integrated with GPS systems to create weed distribution maps for site-specific management.

2. Variable Rate Herbicide Application

Sensors enable real-time weed detection, which can be used to guide variable rate herbicide sprayers. Optical sensors, such as the WeedSeeker® system, detect weeds in real-time and apply herbicides only where needed, reducing chemical usage and environmental impact.

3. Autonomous Weed Control

Robotic weeders equipped with sensors can autonomously navigate fields, detect weeds, and mechanically remove or spray them. For instance, the Naïo Dino robot uses LiDAR and camera-based sensors to identify and target weeds with high precision.

4. Crop-weed Differentiation

Advanced algorithms combined with sensor data can distinguish between crops and weeds, even in complex field conditions. This is particularly useful in row crops where weeds often grow close to the crop canopy.

Recent Innovations in Sensor Technology

Recent advancements have focused on improving sensor accuracy, affordability, and integration with digital platforms. Key innovations include:

- ☆ **AI-Powered Sensors**: Artificial intelligence (AI) algorithms are increasingly being integrated with sensors to enhance weed detection accuracy. For example, AI models trained on large datasets of weed images can process sensor data to identify weed species with high precision.
- ☆ **Miniaturization and Cost Reduction**: Advances in microelectronics have led to the development of compact, low-cost sensors that can be easily mounted on drones or farm machinery.
- ☆ **Multi-Sensor Fusion**: Combining data from multiple sensors, such as optical and thermal sensors, provides a more comprehensive understanding of weed infestations. This approach is particularly effective in heterogeneous field conditions.

Challenges and Future Directions

While sensor technology has made significant strides, challenges remain in terms of cost, scalability, and data processing. Future research is focused on developing more robust sensors that can operate under diverse environmental conditions and integrating them with real-time decision-making platforms.

I. Future of Sensor-Based Weed Control Systems

Sensor-based weed control systems represent a transformative innovation in modern agriculture, offering precision, efficiency, and sustainability in weed management. These systems leverage advanced technologies such as optical sensors, machine vision, and artificial intelligence (AI) to detect, identify, and manage weeds with unprecedented accuracy. As the agricultural industry continues to embrace digitalization, the future of sensor-based weed control systems holds immense potential for revolutionizing weed management practices.

Emerging Technologies in Sensor-Based Weed Control

1. Optical and Multispectral Sensors

Optical sensors, including hyperspectral and multispectral imaging systems, are becoming increasingly sophisticated in detecting weeds based on their unique spectral signatures. These sensors can differentiate between crops and weeds by analyzing light reflectance in specific wavelengths, enabling targeted herbicide application or mechanical removal. For example, hyperspectral imaging can identify weed species even in early growth stages, reducing competition with crops and improving yield potential.

2. AI and Machine Learning Integration

Artificial intelligence (AI) and machine learning algorithms are critical for processing the vast amounts of data generated by sensor systems. These technologies enable real-time weed identification and decision-making by analyzing patterns, shapes, and colors of plants. Machine learning models, trained on extensive datasets, can achieve high accuracy in distinguishing weeds from crops, even under varying environmental conditions (Slaughter *et al.*, 2008).

3. LiDAR and 3D Imaging

LiDAR (Light Detection and Ranging) technology and 3D imaging systems are emerging tools for mapping weed distribution in fields. These systems provide detailed spatial data, allowing for site-specific weed management. LiDAR can also assess plant height and density, offering additional parameters for weed-crop discrimination.

Advantages of Sensor-Based Weed Control

- ☆ **Reduced Herbicide Use**: By enabling site-specific application, these systems minimize herbicide wastage, reducing environmental contamination and input costs.
- ☆ **Labor Efficiency**: Automated weed detection and control reduce the need for manual labor, addressing labor shortages in agriculture.
- ☆ **Sustainability**: Precision weed management supports sustainable farming practices by preserving soil health and biodiversity.

Challenges and Future Directions

Despite their potential, sensor-based weed control systems face challenges such as high initial costs, the need for robust algorithms to handle complex field conditions, and integration with existing farm machinery. Future advancements are likely to focus on improving sensor accuracy, reducing costs, and developing user-friendly interfaces for farmers. Additionally, the integration of Internet of Things (IoT) devices and cloud computing could enable seamless data sharing and remote monitoring, further enhancing the efficiency of these systems.

Applications in Autonomous Weed Control

Autonomous robotic weeders equipped with sensor-based systems are a promising development. These robots can navigate fields independently, detect weeds, and apply control measures with precision. For instance, the "See and Spray" technology by Blue River Technology uses AI-driven cameras to identify and spray weeds selectively, reducing herbicide use by up to 90 per cent (Blue River Technology, 2023).

Chapter 8

Genetic Engineering for Weed Control

Genetic engineering has emerged as a transformative tool in modern weed management, offering precise and sustainable solutions to combat weed infestations. By manipulating the genetic makeup of crops or weeds, scientists can develop innovative strategies to enhance weed control while minimizing environmental impacts.

1. Herbicide-Tolerant Crops (HTC)

One of the most prominent applications of genetic engineering is the development of herbicide-tolerant crops. These crops are genetically modified to withstand specific herbicides, allowing farmers to apply chemical treatments that selectively eliminate weeds without harming the crop. For example:

- ☆ **Glyphosate-tolerant crops** (*e.g.*, soybean, maize) enable the use of glyphosate for broad-spectrum weed control.
- ☆ **Glufosinate-resistant crops** provide similar benefits with alternative herbicide chemistry.

This approach reduces the need for mechanical weeding and promotes conservation tillage, which helps preserve soil health (Duke, 2018).

2. Gene Drives for Weed Suppression

Gene drive technology is an emerging genetic engineering tool designed to alter weed populations. By introducing genetic modifications that disrupt weed reproduction or growth, scientists aim to reduce weed prevalence over time. For instance, gene drives targeting seed viability in invasive species like *Amaranthus palmeri* show promise in reducing its spread (Neve *et al.*, 2018).

3. RNA Interference (RNAi)

RNAi technology offers a novel method for weed control by silencing specific genes essential for weed survival. RNAi-based sprays or seed treatments can target weeds without affecting non-target organisms, making it an environmentally friendly option.

4. Challenges and Considerations

While genetic engineering provides powerful tools, challenges such as the evolution of herbicide-resistant weeds, regulatory hurdles, and public acceptance must be addressed. Additionally, ethical considerations and ecological risks, such as gene flow to wild relatives, require careful evaluation.

A. Introduction to Genetic Engineering for Weed Control

Genetic engineering has emerged as a transformative tool in modern agriculture, offering innovative solutions for weed management. By leveraging advancements in molecular biology, genetic engineering enables precise manipulation of plant genomes to develop crops with enhanced resistance to herbicides, reduced weed competitiveness, or novel mechanisms to suppress weed growth. This approach is particularly valuable in addressing the limitations of conventional weed control methods, such as herbicide resistance and environmental concerns.

Herbicide-Tolerant Crops

One of the most significant applications of genetic engineering in weed control is the development of herbicide-tolerant (HT) crops. These crops are engineered to withstand specific herbicides, allowing farmers to apply broad-spectrum herbicides without damaging the crop. For instance, glyphosate-tolerant crops, such as Roundup Ready soybean and maize, have been widely adopted globally. Glyphosate, a non-selective herbicide, effectively controls a broad range of weeds, reducing the need for multiple herbicide applications and simplifying weed management strategies (Duke and Powles, 2008).

Gene Editing for Weed Suppression

The advent of CRISPR-Cas9 and other gene-editing technologies has opened new possibilities for targeted weed control. Scientists are exploring ways to edit genes in crops to enhance their ability to outcompete weeds or to modify weed genomes to disrupt their growth and reproduction. For example, researchers are investigating the potential of gene drives to spread deleterious traits within weed populations, thereby reducing their fitness and prevalence in agricultural systems.

RNA Interference (RNAi) for Weed Control

RNA interference (RNAi) is another promising genetic engineering tool for weed management. By designing RNA molecules that specifically target essential genes in weeds, RNAi can inhibit their expression, leading to reduced weed growth or viability. This approach offers a highly specific and environmentally friendly alternative to chemical herbicides, as it minimizes off-target effects on non-target organisms.

Challenges and Considerations

While genetic engineering holds great promise for weed control, it also presents challenges. The evolution of herbicide-resistant weeds due to over-reliance on HT crops underscores the need for integrated weed management strategies. Additionally, regulatory hurdles, public perception, and ethical considerations must be addressed to ensure the sustainable adoption of genetically engineered technologies in agriculture.

B. Genetically Modified Crops for Weed Suppression

Genetically Modified (GM) crops have emerged as a transformative tool in modern agriculture, offering innovative solutions for weed management. By integrating specific genetic traits into crops, scientists have developed varieties that can suppress weed growth, reduce herbicide dependency, and enhance crop productivity. This section explores the role of GM crops in weed suppression, the underlying scientific principles, and their practical applications in agriculture.

1. Overview of Genetically Modified Crops in Weed Management

Genetically Modified crops are developed by altering the plant's DNA to introduce desirable traits, such as herbicide tolerance or pest resistance. In the context of weed management, GM crops primarily focus on two approaches:

- ☆ **Herbicide-Tolerant (HT) Crops**: These crops are engineered to withstand specific herbicides, allowing farmers to control weeds effectively without harming the crop.
- ☆ **Allelopathic Crops**: Some GM crops are designed to release natural biochemicals (allelochemicals) into the soil, inhibiting weed germination and growth.

The adoption of GM crops has revolutionized weed control practices, particularly in large-scale farming systems, by simplifying weed management and reducing labor-intensive practices like manual weeding.

2. Herbicide-Tolerant Crops

Mechanism of Action

Herbicide-tolerant crops are engineered to resist broad-spectrum herbicides, such as glyphosate or glufosinate. These herbicides target specific enzymes in plants that are essential for their growth. For example, glyphosate inhibits the enzyme 5-enolpyruvylshikimate-3-phosphate synthase (EPSPS), which is crucial for the synthesis of aromatic amino acids in plants. HT crops are modified to express a glyphosate-resistant version of EPSPS, enabling them to survive herbicide application while weeds are eliminated.

Examples of HT Crops

- **Glyphosate-Tolerant Soybean (GTS 40-3-2)**: Widely adopted in the United States and Brazil, this GM soybean allows for effective glyphosate-based weed control.
- **Roundup Ready Corn**: Engineered to tolerate glyphosate, this corn variety simplifies weed management in maize production systems.

Benefits

- **Selective Weed Control**: HT crops enable the use of non-selective herbicides to target a wide range of weeds.
- **Reduced Tillage**: Farmers can adopt conservation tillage practices, which minimize soil erosion and improve soil health.
- **Labor and Cost Efficiency**: Simplified weed control reduces labor costs and herbicide application frequency.

Challenges

- **Herbicide Resistance in Weeds**: Over-reliance on a single herbicide, such as glyphosate, has led to the emergence of herbicide-resistant weed species like *Amaranthus palmeri* (Palmer amaranth) and *Conyza canadensis* (horseweed) (Heap, 2023).
- **Environmental Concerns**: Increased herbicide use may lead to soil and water contamination.

3. Allelopathic GM Crops

Concept of Allelopathy

Allelopathy refers to the ability of plants to release biochemicals that suppress the growth of neighboring plants, including weeds. Scientists are exploring genetic modifications to enhance the allelopathic properties of crops, enabling them to naturally inhibit weed growth.

Research and Development

- **Rice**: Researchers have identified genes responsible for allelochemical production in rice, such as momilactones and phenolic acids, and are working to enhance their expression through genetic engineering.
- **Wheat**: Efforts are underway to develop wheat varieties with enhanced allelopathic potential to suppress grassy weeds like *Avena fatua* (wild oats).

Advantages

- **Reduced Chemical Dependency**: Allelopathic crops can minimize the need for synthetic herbicides.
- **Sustainable Weed Control**: These crops provide an eco-friendly alternative for managing weeds in integrated farming systems.

Limitations

- **Limited Commercialization**: Allelopathic GM crops are still in the experimental stage and are not widely available for commercial use.
- **Complex Interactions**: The effectiveness of allelopathy depends on environmental factors, soil conditions, and the weed species present.

4. Integrated Weed Management with GM Crops

Role in Integrated Weed Management (IWM)

Genetically Modified crops are a valuable component of Integrated Weed Management (IWM) strategies. By combining GM crops with other weed control methods, such as crop rotation, cover cropping, and mechanical weeding, farmers can achieve sustainable and effective weed suppression.

Best Practices

1. **Herbicide Rotation**: To prevent herbicide resistance, alternate between different herbicide modes of action.
2. **Crop Diversification**: Rotate GM crops with non-GM varieties to disrupt weed life cycles.
3. **Cover Crops**: Use cover crops to reduce weed seed banks and complement the weed-suppressing effects of GM crops.

Case Study

In the United States, the adoption of glyphosate-tolerant soybean in conjunction with conservation tillage and crop rotation has significantly reduced weed pressure and improved soil health. However, the emergence of glyphosate-resistant weeds has necessitated the integration of additional weed control measures (Duke and Powles, 2023).

5. Regulatory and Ethical Considerations

The development and commercialization of GM crops are subject to rigorous regulatory frameworks to ensure their safety for human health and the environment. Key considerations include:

- **Biosafety Assessments**: Evaluating the potential risks of GM crops to non-target organisms and ecosystems.

- ☆ **Public Perception**: Addressing concerns about the long-term impacts of GM crops on biodiversity and food security.
- ☆ **Labeling and Transparency**: Ensuring clear labeling of GM products to provide consumers with informed choices.

6. Future Prospects

Advancements in genetic engineering technologies, such as CRISPR-Cas9, hold promise for developing next-generation GM crops with enhanced weed-suppressing traits. For instance, researchers are exploring gene-editing techniques to improve the precision and efficiency of introducing allelopathic traits into crops.

C. Herbicide-Resistant Crops and their Role in Weed Control

Herbicide-resistant (HR) crops have revolutionized modern weed management, offering a powerful tool to combat weed infestations in agricultural systems. These crops are genetically engineered or selectively bred to tolerate specific herbicides, allowing farmers to apply herbicides that effectively control weeds without damaging the crop. This section explores the scientific principles, benefits, limitations, and practical applications of HR crops in weed control.

Development of Herbicide-Resistant Crops

Herbicide-resistant crops are developed through two primary methods:

1. **Genetic Engineering**: Scientists introduce specific genes into crops to confer resistance to a particular herbicide. For example, glyphosate-resistant crops, such as Roundup Ready soybeans, are engineered to tolerate glyphosate, a broad-spectrum herbicide.
2. **Conventional Breeding**: Resistance traits are developed by selecting and breeding plants with natural tolerance to certain herbicides. This method is less common but has been used for crops like imidazolinone-resistant wheat.

The most widely adopted HR crops include soybeans, maize, cotton, and canola, which are resistant to herbicides such as glyphosate, glufosinate, and dicamba.

Mechanism of Action

Herbicide-resistant crops function by altering the biochemical pathways targeted by specific herbicides. For example:

- ☆ **Glyphosate Resistance**: Glyphosate inhibits the enzyme EPSPS (5-enolpyruvylshikimate-3-phosphate synthase), which is essential for amino acid synthesis in plants. HR crops contain a modified EPSPS enzyme that is unaffected by glyphosate, allowing the crop to survive herbicide application.

- ☆ **Glufosinate Resistance**: Glufosinate inhibits glutamine synthetase, leading to ammonia accumulation and plant death. HR crops express an enzyme that detoxifies glufosinate, ensuring crop survival.

These mechanisms enable selective weed control, where herbicides target weeds while sparing the crop.

Benefits of Herbicide-Resistant Crops

The adoption of HR crops offers several advantages in weed management:

- ☆ **Enhanced Weed Control**: HR crops allow the use of broad-spectrum herbicides, effectively controlling a wide range of weed species.
- ☆ **Simplified Management**: Farmers can apply herbicides post-emergence, reducing the need for pre-emergence herbicides and mechanical weeding.
- ☆ **Reduced Tillage**: HR crops promote conservation tillage practices, minimizing soil erosion and improving soil health.
- ☆ **Increased Yield Potential**: By reducing weed competition, HR crops can enhance crop yields and profitability.

For instance, glyphosate-resistant soybeans have been widely adopted in the United States, leading to significant reductions in weed pressure and increased farm efficiency (Duke and Powles, 2008).

Challenges and Limitations

Despite their advantages, HR crops are not without challenges such as:

1. **Herbicide Resistance in Weeds**: Over-reliance on a single herbicide, such as glyphosate, has led to the evolution of herbicide-resistant weed populations. Examples include glyphosate-resistant Palmer amaranth (Amaranthus palmeri) and waterhemp (Amaranthus tuberculatus).
2. **Environmental Concerns**: Increased herbicide use can lead to off-target effects, such as herbicide drift and contamination of water bodies.
3. **Economic Costs**: HR seeds are often more expensive than conventional seeds, increasing production costs for farmers.
4. **Regulatory and Public Perception Issues**: The use of genetically modified crops is subject to regulatory scrutiny and public debate, which can limit adoption in certain regions.

Best Practices for Sustainable Use

To maximize the benefits of HR crops while mitigating risks, the following best practices are recommended:

- ☆ **Integrated Weed Management (IWM)**: Combine HR crops with other weed control methods, such as crop rotation, cover cropping, and mechanical weeding, to reduce herbicide reliance.
- ☆ **Herbicide Rotation**: Use herbicides with different modes of action to prevent the development of resistant weed populations.
- ☆ **Scouting and Monitoring**: Regularly monitor fields for signs of herbicide-resistant weeds and adjust management strategies accordingly.
- ☆ **Adherence to Label Instructions**: Follow herbicide application guidelines to minimize environmental impacts and ensure efficacy.

Future Prospects

Advancements in biotechnology continue to expand the potential of HR crops. Newer generations of crops are being developed with resistance to multiple herbicides, enabling more flexible and effective weed control strategies. Additionally, research into gene-editing technologies, such as CRISPR-Cas9, holds promise for creating HR crops with improved traits and reduced environmental impact (Zhang *et al.*, 2020).

D. Gene Editing Technologies in Weed Management

Gene editing technologies have emerged as a transformative tool in modern agriculture, offering innovative solutions to the persistent challenge of weed management. These technologies, particularly CRISPR-Cas9 and related systems, enable precise modifications to the genetic material of plants, including weeds, to control their growth, improve crop competitiveness, and reduce reliance on chemical herbicides. This section explores the applications, mechanisms, and potential of gene editing in weed management.

1. Overview of Gene Editing Technologies

Gene editing refers to a suite of molecular techniques that allow targeted modifications to an organism's DNA. The most widely used and advanced tool in this domain is CRISPR-Cas9 (Clustered Regularly Interspaced Short Palindromic Repeats-associated protein 9). Other tools include TALENs (Transcription Activator-Like Effector Nucleases) and ZFNs (Zinc Finger Nucleases). These technologies work by introducing site-specific double-strand breaks in DNA, which are then repaired by the cell's natural mechanisms, leading to precise genetic changes.

Key features of gene editing technologies include:

- ☆ **Precision:** Ability to target specific genes with minimal off-target effects.
- ☆ **Versatility:** Applicable to a wide range of organisms, including weeds.

- ☆ **Efficiency:** Faster and more cost-effective compared to traditional breeding or mutagenesis techniques.

2. Applications in Weed Management

Gene editing technologies hold immense potential for addressing weed-related challenges in agriculture. Some key applications include:

a. Herbicide Resistance Management

Weeds often develop resistance to commonly used herbicides, rendering chemical control methods ineffective. Gene editing can be employed to:

- ☆ **Reverse Herbicide Resistance:** Modify the genetic mutations in weeds that confer resistance to herbicides, restoring their susceptibility.
- ☆ **Enhance Crop Tolerance:** Engineer crops to tolerate specific herbicides, enabling selective weed control without harming the crop.

For example, CRISPR-Cas9 has been used to target the EPSPS (5-enolpyruvylshikimate-3-phosphate synthase) gene in weeds, which is commonly associated with glyphosate resistance (Li *et al.*, 2022).

b. Sterility Induction in Weeds

Gene editing can be utilized to induce sterility in problematic weed species by disrupting genes involved in reproduction. This approach reduces the weed seed bank in the soil over time, leading to long-term weed population control.

c. Altering Weed Growth Characteristics

By targeting genes responsible for growth and development, gene editing can be used to slow down weed growth, reduce seed production, or make weeds more susceptible to environmental stresses.

d. Biocontrol Enhancement

Gene editing can improve the efficacy of biological control agents, such as fungi or bacteria, by enhancing their ability to target specific weeds.

3. Mechanisms of Gene Editing in Weeds

The application of gene editing in weed management involves several steps:

1. **Target Gene Identification:** Identifying genes responsible for herbicide resistance, reproduction, or other undesirable traits in weeds.
2. **Guide RNA Design:** Designing guide RNAs (gRNAs) to direct the CRISPR-Cas9 system to the target gene.
3. **Delivery System:** Introducing the gene-editing machinery into weed cells using methods such as particle bombardment, electroporation, or Agrobacterium-mediated transformation.

4. **Validation:** Confirming the desired genetic modification and assessing its impact on weed physiology and behavior.

4. Challenges and Limitations

While gene editing offers significant promise, its application in weed management is not without challenges:

- **Off-Target Effects:** Although CRISPR-Cas9 is highly precise, unintended genetic modifications can occur, potentially leading to undesirable traits.
- **Regulatory Hurdles:** The use of gene-edited organisms in agriculture is subject to stringent regulatory frameworks, which vary across countries.
- **Ethical Concerns:** Public perception and ethical considerations surrounding genetic modification may hinder widespread adoption.
- **Delivery Efficiency:** Efficient delivery of gene-editing tools into weed species remains a technical challenge, particularly for species with complex genomes.

5. Future Prospects

Advances in gene editing technologies, such as base editing and prime editing, are expected to further enhance the precision and efficiency of weed management strategies. Additionally, the integration of gene editing with other technologies, such as artificial intelligence and high-throughput phenotyping, could accelerate the identification of target genes and the development of novel weed control solutions.

E. Environmental and Ecological Impacts of Genetic Engineering

Genetic engineering (GE) has revolutionized modern agriculture, particularly in weed management, by enabling the development of herbicide-tolerant (HT) crops and other genetically modified organisms (GMOs). While these technologies have contributed to increased agricultural productivity, their environmental and ecological implications warrant careful consideration. This section explores the potential impacts of genetic engineering on ecosystems, biodiversity, and environmental sustainability.

1. Gene Flow and Non-Target Effects

One of the most significant ecological concerns associated with genetic engineering is the unintended transfer of genes from genetically modified crops to wild relatives or non-target species. This phenomenon, known as gene flow, can have far-reaching consequences:

- ☆ **Development of Herbicide-Resistant Weeds**: The widespread adoption of HT crops, such as glyphosate-tolerant soybeans and maize, has led to the emergence of herbicide-resistant weed populations. For instance, species like *Amaranthus palmeri* (Palmer amaranth) have evolved resistance through both natural selection and potential gene flow from HT crops (Heap, 2023). These "superweeds" pose significant challenges to weed management and necessitate the use of more aggressive chemical or mechanical control methods.
- ☆ **Impact on Wild Plant Populations**: Gene flow can introduce transgenes into wild plant populations, potentially altering their genetic makeup and ecological roles. For example, the transfer of herbicide-tolerance genes to wild relatives could lead to the proliferation of invasive species, disrupting native ecosystems.

2. Biodiversity Loss

The adoption of genetically engineered crops has raised concerns about its impact on biodiversity at multiple levels:

- ☆ **Reduction in Crop Diversity**: The dominance of a few genetically engineered crop varieties, such as glyphosate-tolerant soybean and Bt cotton, has led to a decline in the cultivation of traditional and diverse crop varieties. This reduction in genetic diversity increases the vulnerability of agricultural systems to pests, diseases, and climate change.
- ☆ **Effects on Non-Target Organisms**: Genetic engineering can indirectly affect non-target organisms, including beneficial insects, soil microorganisms, and wildlife. For instance:
- ☆ Bt crops, engineered to produce insecticidal proteins derived from *Bacillus thuringiensis*, may inadvertently harm non-target insects such as pollinators and natural pest predators (Losey *et al.*, 1999).
- ☆ Changes in weed populations due to herbicide-tolerant crops can alter habitat availability for insects, birds, and other wildlife that depend on diverse weed flora.

3. Soil Health and Microbial Communities

The environmental impacts of genetic engineering extend to soil ecosystems, which are critical for sustainable agriculture:

- ☆ **Herbicide Overuse**: The reliance on herbicide-tolerant crops has led to increased herbicide application, particularly glyphosate. Prolonged and excessive use of glyphosate can disrupt soil microbial communities, reducing microbial diversity and altering nutrient cycling processes. Studies have shown that glyphosate residues can inhibit the growth of beneficial soil fungi and bacteria, potentially affecting soil fertility and plant health (Zaller *et al.*, 2018).

- ☆ **Impact on Soil Biodiversity**: Transgenic crops may release novel compounds, such as Bt toxins, into the soil through root exudates or decaying plant material. These compounds can accumulate in the soil and affect non-target soil organisms, including earthworms and nematodes, which play essential roles in maintaining soil structure and fertility.

4. Herbicide Resistance and Escalating Chemical Use

The introduction of HT crops was initially intended to reduce herbicide use by enabling targeted weed control. However, the emergence of herbicide-resistant weeds has led to a counterproductive trend:

- ☆ **Increased Herbicide Application**: Farmers often resort to applying higher doses of herbicides or using multiple herbicide formulations to combat resistant weeds. This practice not only increases production costs but also exacerbates environmental contamination and the risk of herbicide runoff into water bodies.
- ☆ **Chemical Pollution**: Herbicide residues can persist in the environment, contaminating soil, water, and air. This pollution poses risks to aquatic ecosystems, where herbicides can affect non-target aquatic plants and organisms, disrupting food chains and ecosystem balance.

5. Long-Term Ecological Risks

The long-term ecological consequences of genetic engineering remain uncertain due to the complexity of ecosystems and the relatively short history of GMO adoption. Potential risks include:

- ☆ **Unpredictable Interactions**: The introduction of transgenes into complex ecosystems may lead to unforeseen interactions between organisms, potentially destabilizing ecological networks.
- ☆ **Evolutionary Pressures**: The widespread use of genetic engineering in agriculture exerts evolutionary pressures on weeds, pests, and pathogens, driving the rapid evolution of resistance mechanisms. This evolutionary arms race can undermine the effectiveness of GE technologies over time.

6. Regulatory and Monitoring Challenges

Effective regulation and monitoring are critical to mitigating the environmental and ecological impacts of genetic engineering. However, challenges persist:

- ☆ **Inadequate Risk Assessment**: Current risk assessment frameworks often fail to account for long-term and cumulative ecological impacts, particularly in diverse and interconnected ecosystems.
- ☆ **Lack of Post-Market Monitoring**: Many countries lack robust systems for monitoring the environmental effects of GMOs after their commercial release, making it difficult to identify and address emerging issues.

F. Benefits of Genetic Engineering in Weed Control

Genetic engineering has revolutionized agriculture, offering innovative solutions to age-old challenges, including weed management. By introducing desirable traits into crops, genetic engineering enables farmers to manage weeds more effectively, sustainably, and economically. This section explores the key benefits of genetic engineering in weed control, supported by scientific principles and real-world applications.

1. Herbicide-Tolerant Crops

One of the most significant advancements in weed control through genetic engineering is the development of herbicide-tolerant (HT) crops. These crops are engineered to withstand specific herbicides, allowing farmers to target weeds without harming the crop itself.

- ☆ **Mechanism**: HT crops are designed by inserting genes that confer resistance to broad-spectrum herbicides, such as glyphosate or glufosinate. For example, glyphosate-tolerant crops (*e.g.*, Roundup Ready® soybeans) contain a modified version of the enzyme 5-enolpyruvylshikimate-3-phosphate synthase (EPSPS), which is unaffected by glyphosate application.

Advantages

- ☆ **Selective Weed Control**: Farmers can apply herbicides post-emergence to eliminate weeds while leaving the crop unharmed.
- ☆ **Reduced Tillage**: HT crops facilitate no-till or reduced-till farming systems, which help conserve soil structure and moisture.
- ☆ **Cost-Effectiveness**: Broad-spectrum herbicides are often cheaper and require fewer applications compared to selective herbicides.

In-text citation: Glyphosate-tolerant crops have been widely adopted globally, with significant reductions in weed pressure and labor costs reported in countries like the United States and Brazil (Duke, 2018).

2. Reduction in Herbicide Use

Genetic engineering has enabled the use of more efficient weed control strategies, often requiring fewer herbicide applications. This not only reduces input costs but also minimizes the environmental impact of chemical use.

- **Target-Specific Herbicides**: HT crops allow for the use of herbicides that are less persistent in the environment, such as glyphosate, which has a relatively short half-life in soil.
- **Lower Dosages**: The precision of genetic engineering ensures that herbicides are applied only when necessary, reducing the overall chemical load on agricultural ecosystems.

Example: Studies have shown that the adoption of HT crops has led to a 37 per cent reduction in herbicide use in soybean production systems in the United States (Brookes and Barfoot, 2020).

3. Improved Weed Resistance Management

Weeds often develop resistance to herbicides over time, rendering traditional chemical control methods less effective. Genetic engineering offers tools to combat this issue by enabling the development of crops with resistance to multiple herbicides.

- **Stacked Traits**: Crops can be engineered to tolerate multiple herbicides, allowing for the rotation or combination of herbicides with different modes of action. This reduces the selection pressure on weeds to develop resistance.
- **Integrated Weed Management (IWM)**: HT crops can be integrated with other weed control methods, such as crop rotation and mechanical weeding, to delay the onset of herbicide resistance.

Scientific Insight: The use of stacked HT traits, such as glyphosate and dicamba tolerance in cotton, has been shown to effectively manage glyphosate-resistant weeds like Palmer amaranth (Amaranthus palmeri) (Heap, 2022).

4. Enhanced Crop Competitiveness

Genetic engineering can improve the competitive ability of crops against weeds by introducing traits that enhance their growth, nutrient uptake, or stress tolerance.

- **Faster Growth Rates**: Crops engineered for rapid early growth can outcompete weeds for light, water, and nutrients.
- **Allelopathy**: Genetic engineering can enhance a crop's ability to release allelopathic compounds, which inhibit weed germination and growth.

Example: Research is ongoing to develop rice varieties with enhanced allelopathic properties to suppress weed species like barnyardgrass (Echinochloa crus-galli).

5. Environmental Benefits

The adoption of genetically engineered crops for weed control has several environmental advantages, contributing to sustainable agriculture.

- **Soil Conservation**: HT crops support no-till farming, which reduces soil erosion and enhances carbon sequestration.
- **Biodiversity Preservation**: Reduced herbicide use and targeted weed control help maintain non-target plant species and beneficial organisms in the ecosystem.
- **Lower Carbon Footprint**: By reducing the need for mechanical weeding and multiple herbicide applications, genetic engineering contributes to lower greenhouse gas emissions from farming operations.

In-text citation: Studies indicate that the adoption of genetically engineered crops has reduced greenhouse gas emissions equivalent to removing 15.3 million cars from the road annually (Brookes and Barfoot, 2020).

6. Economic Benefits for Farmers

Genetic engineering in weed control offers significant economic advantages, particularly for smallholder farmers in developing countries.

- **Higher Yields**: Effective weed control reduces competition for resources, leading to increased crop yields.
- **Labor Savings**: HT crops reduce the need for manual weeding, which is labor-intensive and costly.
- **Market Access**: Genetically engineered crops often have higher market value due to their consistent quality and yield.

Example: In India, the adoption of genetically engineered Bt cotton has not only improved pest resistance but also reduced weed pressure, leading to higher profitability for farmers.

7. Future Potential

The field of genetic engineering continues to evolve, offering exciting possibilities for weed control in the future.

- **CRISPR-Cas9 Technology**: Gene-editing tools like CRISPR-Cas9 are being explored to develop crops with enhanced weed resistance traits, such as improved allelopathy or faster growth rates.
- **Weed-Specific Genetic Targets**: Research is underway to use genetic engineering to directly target and suppress weed species without affecting crops or non-target plants.

Scientific Insight: CRISPR technology has been successfully used to modify rice genes to enhance resistance to parasitic weeds like Striga hermonthica (Kountche *et al.*, 2022).

G. Challenges and Ethical Considerations for Genetic Engineering for Weed Control

Genetic engineering (GE) has emerged as a promising tool in modern weed management, offering innovative solutions to mitigate weed infestations. However, its application is accompanied by significant challenges and ethical considerations that must be addressed to ensure sustainable and responsible use. This section explores these aspects, focusing on technical, ecological, socio-economic, and ethical dimensions.

Challenges in Genetic Engineering for Weed Control

1. **Development of Herbicide-Resistant Weeds:** One of the primary challenges associated with GE crops is the unintended evolution of herbicide-resistant weed populations. For instance, the widespread adoption of glyphosate-resistant crops has led to the proliferation of glyphosate-resistant weeds, such as *Amaranthus palmeri* (Palmer amaranth) and *Lolium rigidum* (rigid ryegrass) (Heap, 2023). This resistance arises from the selective pressure exerted by the overuse of herbicides, necessitating the development of alternative strategies.
2. **Gene Flow and Cross-Pollination:** The escape of transgenes from genetically engineered crops to wild relatives or non-GE crops through cross-pollination poses ecological risks. For example, the transfer of herbicide-resistance genes to wild weed species could create "superweeds" that are difficult to control. This gene flow can occur via pollen dispersal, seed movement, or hybridization, complicating weed management efforts.
3. **Unintended Ecological Consequences:** Genetic modifications may have unintended effects on non-target organisms and ecosystems. For instance, herbicide-tolerant crops might alter soil microbial communities or impact beneficial insects. Additionally, the reduction of weed diversity in agricultural fields could negatively affect biodiversity and ecosystem services, such as pollination and pest control.
4. **Regulatory and Technical Hurdles:** The development and commercialization of genetically engineered crops for weed control are subject to stringent regulatory frameworks. These processes are time-consuming and costly, often requiring extensive testing for environmental safety, human health impacts, and efficacy. Furthermore, the technical complexity of engineering traits such as weed-specific lethality or sterility adds to the challenges.
5. **Public Perception and Acceptance:** Public skepticism and resistance to genetically modified organisms (GMOs) remain significant barriers. Concerns about food safety, environmental risks, and corporate control over agricultural technologies have fueled opposition to GE crops in

many regions. This resistance can hinder the adoption of GE-based weed control strategies, despite their potential benefits.

Ethical Considerations in Genetic Engineering for Weed Control

1. **Equity and Access:** The development and deployment of GE technologies are often dominated by large multinational corporations, raising concerns about equitable access for smallholder farmers. High costs associated with GE seeds and dependence on proprietary technologies may exacerbate socio-economic disparities in agriculture.
2. **Environmental Stewardship:** Ethical considerations demand a precautionary approach to the environmental impacts of GE crops. Scientists and policymakers must weigh the potential benefits of reduced herbicide use against the risks of unintended ecological harm. Long-term monitoring and adaptive management are essential to uphold environmental integrity.
3. **Informed Consent and Transparency:** Farmers and consumers have the right to make informed decisions about the use of GE crops. Transparent labeling, clear communication of risks and benefits, and stakeholder engagement are critical to fostering trust and ensuring ethical deployment of these technologies.
4. **Intergenerational Responsibility:** The long-term consequences of genetic engineering for weed control extend beyond the present generation. Ethical stewardship requires consideration of how current decisions will impact future agricultural systems, biodiversity, and ecological resilience.

Conclusion

The challenges and ethical considerations surrounding genetic engineering for weed control highlight the need for a balanced, science-based approach. Addressing these issues requires collaboration among researchers, policymakers, industry stakeholders, and the public to develop sustainable and socially acceptable solutions.

H. Regulations and Public Perception of Genetically Engineered Weeds

The advent of genetic engineering in agriculture has introduced innovative tools for weed management, including the potential development of genetically engineered (GE) weeds. While these technologies offer promising solutions to persistent weed challenges, they also raise complex regulatory and public perception issues. This section explores the regulatory frameworks governing GE weeds and the societal attitudes shaping their acceptance.

Regulatory Frameworks for Genetically Engineered Weeds

The regulation of genetically engineered organisms, including weeds, is a critical aspect of ensuring environmental safety and public trust. Regulatory frameworks vary across countries but generally focus on assessing risks related to ecological balance, gene flow, and unintended consequences.

1. **United States**: In the U.S., the regulation of GE organisms falls under the coordinated framework of the USDA, EPA, and FDA. For GE weeds, the USDA's Animal and Plant Health Inspection Service (APHIS) evaluates potential impacts on plant health, while the EPA assesses environmental risks, such as gene transfer to non-target species. The FDA ensures food safety if the GE weed has any implications for human consumption.
2. **European Union**: The EU adopts a precautionary principle in regulating genetically modified organisms (GMOs), including GE weeds. The European Food Safety Authority (EFSA) conducts rigorous risk assessments, focusing on biodiversity, ecosystem stability, and potential cross-breeding with native flora. Public consultation is a mandatory part of the approval process, reflecting the EU's emphasis on transparency.
3. **International Guidelines**: The Cartagena Protocol on Biosafety provides an international framework for the safe transfer, handling, and use of living modified organisms (LMOs), including GE weeds. It emphasizes risk assessment, monitoring, and public awareness to mitigate potential adverse effects on biodiversity.

Key Regulatory Challenges

- **Gene Flow and Hybridization**: One of the primary concerns is the potential for GE weeds to crossbreed with wild relatives, leading to the unintended spread of engineered traits, such as herbicide resistance, into natural ecosystems.
- **Ecosystem Disruption**: Regulatory bodies must evaluate whether GE weeds could outcompete native species, disrupt food chains, or alter soil microbiota.
- **Long-Term Monitoring**: Many regulatory frameworks lack robust mechanisms for long-term monitoring of GE weeds post-release, making it difficult to assess cumulative impacts over time.

Public Perception of Genetically Engineered Weeds

Public attitudes toward GE weeds are shaped by a combination of scientific understanding, cultural values, and trust in regulatory institutions. While some stakeholders recognize the potential benefits of GE weeds, such as reduced herbicide use and improved weed control, others express concerns about ecological risks and ethical considerations.

Factors Influencing Public Perception

- **Risk Perception**: The public often perceives GE weeds as high-risk due to uncertainties surrounding their ecological impacts. This perception is amplified by media coverage of worst-case scenarios, such as the creation of "superweeds."
- **Trust in Science and Regulation**: Public confidence in the scientific community and regulatory agencies plays a significant role in shaping attitudes. Transparent communication and stakeholder engagement are essential to building trust.
- **Cultural and Ethical Concerns**: In some regions, genetic engineering is viewed as an unnatural interference with nature, leading to resistance against the adoption of GE technologies.

Examples of Public Response

- In the EU, public opposition to GMOs, including GE weeds, has led to stringent regulatory requirements and limited adoption.
- In contrast, countries like Brazil and Argentina have seen greater acceptance of genetically engineered crops, partly due to strong agricultural lobbies and public awareness campaigns highlighting economic benefits.

Balancing Innovation and Public Concerns

The development of GE weeds requires a delicate balance between harnessing their potential benefits and addressing public concerns. Regulatory agencies must ensure rigorous risk assessments while fostering transparent communication to bridge the gap between scientific advancements and societal acceptance.

I. Future Directions in Genetic Engineering for Weed Control

Genetic engineering holds immense potential for revolutionizing weed management by offering precise, sustainable, and environmentally friendly solutions. Advances in molecular biology and biotechnology are paving the way for innovative approaches to control weeds with minimal impact on non-target organisms and ecosystems. Below are some of the most promising future directions in this field.

1. Development of Herbicide-Resistant Crops

- The introduction of herbicide-resistant (HR) crops has already transformed weed management by enabling the use of broad-spectrum herbicides like glyphosate. Future efforts are focused on:

- Engineering crops with resistance to multiple herbicides, allowing for the rotation of herbicides to mitigate resistance development in weeds.
- Developing resistance to newer, safer herbicides with improved environmental profiles.
- For example, CRISPR-Cas9 technology is being explored to create HR crops with higher precision, reducing the risk of unintended genetic modifications (Chen *et al.*, 2022).

2. Gene Drives for Weed Suppression

- Gene drive systems, which bias the inheritance of specific genes, are being investigated to suppress weed populations. Potential applications include:
- Spreading genes that reduce weed fertility or viability.
- Targeting invasive weed species to curb their spread without affecting native flora.
- This approach is particularly promising for controlling problematic weeds like Palmer amaranth (Amaranthus palmeri), which has developed resistance to multiple herbicides.

3. RNA Interference (RNAi) for Targeted Weed Control

- RNA interference (RNAi) is a gene-silencing technology that can be used to disrupt essential genes in weeds. Future directions include:
- Developing RNAi-based sprays that specifically target weed species without harming crops or beneficial organisms.
- Engineering crops to produce RNAi molecules that affect nearby weeds, creating a self-protecting system.
- For instance, RNAi has been successfully used to target genes involved in seed germination in weeds, offering a novel pre-emergence control strategy (Zhang *et al.*, 2021).

4. Synthetic Biology for Custom Weed Management Solutions

- Synthetic biology enables the design of entirely new biological systems. Potential applications in weed control include:
- Engineering soil microbes to produce natural herbicides or compounds that inhibit weed growth.
- Designing crops with enhanced allelopathic properties to suppress weeds naturally.
- These approaches could reduce reliance on synthetic chemical herbicides, promoting sustainable agriculture.

5. Precision Weed Management Using Genetic Tools

- ☆ Advances in genetic engineering are being integrated with precision agriculture technologies. Examples include:
- ☆ Developing biosensors in crops that detect weed presence and trigger localized herbicide production.
- ☆ Engineering crops with traits that make them more competitive against weeds in specific environments.

Challenges and Ethical Considerations

- ☆ While genetic engineering offers transformative potential, challenges such as regulatory hurdles, public acceptance, and the risk of gene flow to non-target species must be addressed. Ethical considerations, including the long-term ecological impact, also require careful evaluation.

Chapter 9

Nano Sensor and Nano Herbicides

Nanotechnology has emerged as a revolutionary tool in modern weed management, offering precision, efficiency, and sustainability. Nano sensors and nano herbicides are two pivotal advancements that leverage nanoscale materials for targeted weed control and monitoring.

Nano Sensors in Weed Management

Nano sensors are highly sensitive devices capable of detecting minute changes in the environment. In weed management, they are used to identify weed species, monitor herbicide application, and assess soil health. These sensors utilize nanoscale materials like carbon nanotubes, quantum dots, or metal nanoparticles to detect specific biochemical markers or environmental cues.

Applications

- Real-time detection of weed infestations using biochemical signatures.
- Monitoring herbicide residues in soil to prevent over-application.
- Assessing soil nutrient levels to optimize crop-weed competition management.

For instance, nano sensors embedded in drones or autonomous vehicles can map weed distribution across fields, enabling site-specific herbicide application (Kumar *et al.*, 2021).

Nano Herbicides

Nano herbicides are herbicidal formulations that incorporate nanoparticles to enhance efficacy, reduce environmental impact, and minimize herbicide resistance. These formulations improve solubility, stability, and targeted delivery of active ingredients.

Advantages

- Controlled release of herbicides for prolonged activity.
- Reduced off-target effects, lowering environmental contamination.
- Enhanced penetration into weed tissues for improved efficacy.

For example, silica nanoparticles loaded with glyphosate have shown increased herbicidal activity against resistant weed species (Jampílek and Králová, 2021).

Challenges and Future Prospects

While promising, the adoption of nano sensors and nano herbicides faces challenges such as high production costs, regulatory hurdles, and potential ecological risks. Ongoing research aims to address these issues and unlock the full potential of nanotechnology in sustainable weed management.

A. Introduction to Nano Sensors and Nano Herbicides

The advent of nanotechnology has revolutionized various sectors, including agriculture, by offering innovative solutions for sustainable weed management. Nano sensors and nano herbicides are two cutting-edge applications of nanotechnology that are gaining traction in modern weed control strategies. These technologies promise precision, efficiency, and environmental sustainability, addressing the limitations of conventional weed management practices.

Nano Sensors in Weed Management

Nano sensors are highly sensitive devices that operate at the nanoscale, capable of detecting and monitoring specific biological, chemical, or physical parameters. In weed management, nano sensors are utilized for real-time detection and monitoring of weed infestations, soil health, and herbicide residues. Their high sensitivity and specificity enable early detection of weeds, which is critical for timely intervention.

Key features of nano sensors in weed management include:

- **Real-Time Monitoring**: Nano sensors can detect weed species based on their unique biochemical markers, such as secondary metabolites or volatile organic compounds (VOCs). For instance, carbon nanotube-based sensors can identify specific weed VOCs with high accuracy.
- **Precision Agriculture**: By integrating nano sensors with GPS and IoT (Internet of Things) technologies, farmers can create weed distribution maps, enabling site-specific herbicide application.

☆ **Environmental Safety**: Nano sensors can monitor herbicide residues in soil and water, ensuring compliance with environmental safety standards.

Recent advancements include the development of biosensors that use enzyme or antibody-based recognition systems to detect weed-specific proteins or DNA sequences (Kumar *et al.*, 2022).

Nano Herbicides: A Targeted Approach

Nano herbicides involve the encapsulation of active herbicidal ingredients within nanoscale carriers, such as liposomes, polymeric nanoparticles, or silica nanoparticles. This encapsulation enhances the efficacy and safety of herbicides by enabling targeted delivery and controlled release.

Advantages of nano herbicides include:

1. **Enhanced Efficacy**: Nano carriers improve the solubility and stability of herbicides, ensuring better absorption by weeds.
2. **Targeted Delivery**: Nano herbicides can be engineered to release their active ingredients only in the presence of specific weed enzymes or pH conditions, minimizing off-target effects.
3. **Reduced Dosage**: Controlled release mechanisms reduce the overall quantity of herbicides required, lowering costs and environmental impact.
4. **Compatibility with Biodegradable Materials**: Many nano carriers are biodegradable, reducing the risk of soil and water contamination.

For example, silica-based nano herbicides have shown promising results in delivering glyphosate with improved efficiency and reduced environmental risks (Singh *et al.*, 2021).

Integration of Nano Sensors and Nano Herbicides

The integration of nano sensors with nano herbicides offers a synergistic approach to weed management. Nano sensors can guide the precise application of nano herbicides, ensuring that only infested areas are treated. This integration reduces herbicide wastage, mitigates resistance development in weeds, and promotes sustainable agricultural practices.

B. What are Nano Sensors?

Nanotechnology has revolutionized various fields of science, including agriculture. Among its many applications, nano sensors have emerged as a promising tool for precision agriculture, including modern weed management. Nano sensors are devices that operate at the nanoscale (1–100 nanometers) and are capable of detecting and responding to physical, chemical, or biological stimuli with high sensitivity and specificity. In the context of weed management, nano sensors offer innovative solutions for early weed detection, herbicide optimization, and environmental monitoring.

Principle of Nano Sensors

Nano sensors function based on the unique properties of nanomaterials, such as high surface area, quantum effects, and enhanced reactivity. These properties enable nano sensors to detect minute changes in their environment, such as the presence of specific molecules, temperature variations, or light intensity. The working mechanism typically involves:

- ☆ **Recognition Element**: A component that interacts with the target analyte (*e.g.*, weed-specific volatile organic compounds or herbicide residues).
- ☆ **Transducer**: Converts the interaction into a measurable signal, such as an electrical, optical, or mechanical response.
- ☆ **Signal Processor**: Amplifies and processes the signal for interpretation.

For example, carbon nanotubes and metal nanoparticles are commonly used in nano sensors due to their exceptional electrical and optical properties (Kumar *et al.*, 2021).

C. Applications of Nano Sensors in Weed Management

Nano sensors have the potential to transform weed management practices by enabling real-time, site-specific interventions. Below are some key applications:

1. Early Weed Detection

- ☆ Nano sensors can detect weed-specific biomarkers, such as volatile organic compounds (VOCs) released by weeds during germination or growth.
- ☆ For instance, nano sensors integrated with drones or autonomous vehicles can scan fields for the presence of weeds and provide geospatial data for targeted interventions.

2. Precision Herbicide Application

- ☆ Nano sensors can monitor the concentration of herbicides in the soil or on plant surfaces to ensure optimal application rates.
- ☆ They can also detect herbicide resistance in weeds by identifying specific genetic or biochemical markers, enabling farmers to adjust their weed control strategies accordingly.

3. Environmental Monitoring

- ☆ Nano sensors can measure herbicide residues in soil and water, ensuring compliance with environmental regulations.

- They can also detect soil nutrient levels and pH, which indirectly influence weed growth and herbicide efficacy.

4. **Stress Monitoring in Crops**

- Nano sensors can differentiate between crop stress caused by weeds and other factors, such as nutrient deficiencies or pest infestations. This helps in prioritizing weed control measures.

Types of Nano Sensors Used in Weed Management

Several types of nano sensors are being explored for agricultural applications, including:

- **Chemical Nano Sensors**: Detect specific chemical compounds, such as herbicide residues or weed-emitted VOCs. For example, gold nanoparticles are used for detecting trace amounts of herbicides in soil.
- **Optical Nano Sensors**: Use light-based detection methods, such as fluorescence or surface plasmon resonance, to identify weed-specific markers.
- **Electrochemical Nano Sensors**: Measure electrical signals generated by the interaction between the sensor and the target analyte. These are particularly useful for detecting herbicide resistance at the molecular level.

Advantages of Nano Sensors in Weed Management

Nano sensors offer several advantages over traditional weed management methods:

- **High Sensitivity and Specificity**: Capable of detecting weeds or herbicide residues at extremely low concentrations.
- **Real-Time Monitoring**: Provide immediate feedback, enabling timely interventions.
- **Reduced Herbicide Use**: Optimize herbicide application, minimizing environmental impact and cost.
- **Scalability**: Can be integrated into existing precision agriculture systems, such as drones, robots, or IoT devices.

Challenges and Future Directions

Despite their potential, the adoption of nano sensors in weed management faces several challenges:

- **Cost**: The high cost of manufacturing nano sensors limits their widespread use.

- ☆ **Durability**: Nano sensors must withstand harsh agricultural environments, including temperature fluctuations and soil moisture.
- ☆ **Regulatory Hurdles**: The use of nanomaterials in agriculture is subject to stringent regulations due to potential environmental and health risks.

Future research is focused on developing cost-effective, durable, and multifunctional nano sensors that can address these challenges. Integration with machine learning and artificial intelligence (AI) is also being explored to enhance the decision-making process in weed management (Singh *et al.*, 2022).

D. Applications of Nano Sensors in Weed Control

The integration of nanotechnology into agriculture has opened new frontiers in precision weed management. Nano sensors, a subset of nanotechnology, are highly sensitive devices capable of detecting and monitoring environmental and biological parameters at the nanoscale. Their application in weed control offers innovative solutions for early detection, precise herbicide application, and real-time monitoring of weed populations. This section explores the role of nano sensors in modern weed management, emphasizing their potential to enhance efficiency and sustainability.

1. Early Detection of Weed Infestations

Nano sensors can detect weeds at their earliest growth stages, enabling timely intervention. These sensors are designed to identify specific biochemical markers, such as volatile organic compounds (VOCs) or specific proteins released by weeds during germination or growth.

- ☆ **Example:** Nano sensors embedded in soil or mounted on drones can detect VOCs emitted by weed seeds during germination. This allows farmers to target weeds before they compete with crops for resources.
- ☆ **Scientific Principle:** Nano sensors utilize surface plasmon resonance (SPR) or fluorescence-based detection to identify weed-specific biomarkers at extremely low concentrations.

Early detection helps reduce the need for broad-spectrum herbicides, minimizing environmental impact and lowering costs.

2. Precision Herbicide Application

Nano sensors play a critical role in site-specific weed management by enabling precise herbicide application. These sensors can differentiate between crop plants and weeds based on their spectral signatures, leaf morphology, or biochemical composition.

- ☆ **Integration with Sprayers:** Nano sensors integrated with smart sprayers can identify weed patches in real-time and apply herbicides only where needed. This reduces chemical usage and prevents damage to non-target plants.

- **Example:** Optical nano sensors mounted on autonomous robots can scan fields and trigger herbicide nozzles to spray only on detected weeds.

By ensuring precise targeting, nano sensors contribute to reducing herbicide resistance in weed populations and promoting sustainable farming practices.

3. Monitoring Herbicide Resistance

Herbicide resistance is a growing challenge in weed management. Nano sensors can detect genetic or biochemical markers associated with herbicide resistance in weed populations.

- **Example:** Nano biosensors can identify mutations in weed DNA that confer resistance to specific herbicides. This information helps agronomists adjust their weed management strategies accordingly.
- **Scientific Principle:** DNA-based nano sensors use complementary DNA (cDNA) probes to bind with specific genetic sequences, enabling the detection of resistance traits.

This capability allows for proactive management of herbicide resistance, ensuring the long-term efficacy of weed control programs.

4. Real-Time Field Monitoring

Nano sensors enable continuous, real-time monitoring of weed populations and environmental conditions. These sensors can be deployed in the field as part of an Internet of Things (IoT) network, providing farmers with actionable data.

- **Example:** Wireless nano sensors embedded in soil can monitor weed density, soil moisture, and nutrient levels simultaneously. This data is transmitted to a central system for analysis, enabling informed decision-making.
- **Benefits:** Real-time monitoring reduces labor costs, improves resource allocation, and enhances the overall efficiency of weed management practices.

5. Environmental Benefits

The use of nano sensors in weed control significantly reduces the environmental footprint of agriculture. By minimizing herbicide usage and preventing over-application, nano sensors help protect soil health, water quality, and biodiversity.

- **Example:** Nano sensors can detect herbicide residues in soil and water, ensuring compliance with environmental regulations and preventing contamination.

E. Advantages of Nano Sensors in Precision Agriculture

Nanotechnology has emerged as a transformative tool in modern agriculture, particularly in precision farming and weed management. Nano sensors, a key application of nanotechnology, are highly sensitive devices capable of detecting minute changes in environmental conditions, soil health, and crop status. Their integration into precision agriculture offers significant advantages, enabling farmers to optimize inputs, reduce costs, and improve sustainability.

1. Enhanced Weed Detection and Monitoring

Nano sensors play a pivotal role in detecting weed infestations at an early stage. By identifying specific biochemical markers or volatile organic compounds (VOCs) released by weeds, nano sensors provide real-time data on weed presence and density. This early detection allows for targeted intervention, reducing the need for blanket herbicide applications.

- ☆ **Example**: Nano sensors embedded in drones or autonomous vehicles can scan large fields and identify weed patches with high accuracy, enabling site-specific weed control.
- ☆ **Scientific Principle**: Nano sensors utilize nanoscale materials like carbon nanotubes or metal oxides, which exhibit high sensitivity to chemical signals, to detect weed-related compounds.

2. Precision Herbicide Application

Nano sensors contribute to the precise application of herbicides by identifying the exact location and extent of weed infestations. This precision reduces herbicide wastage, minimizes environmental contamination, and lowers production costs.

- ☆ **Key Advantage**: Farmers can apply herbicides only where necessary, avoiding overuse and reducing the risk of herbicide resistance in weeds.
- ☆ **Example**: Nano-enabled smart sprayers equipped with sensors can differentiate between crops and weeds, ensuring selective herbicide application.

3. Real-Time Soil Health Monitoring

Weed growth is often influenced by soil conditions such as nutrient levels, pH, and moisture. Nano sensors provide continuous, real-time monitoring of these parameters, helping farmers understand the factors contributing to weed proliferation.

- ☆ **Benefits**:
- ☆ Early identification of soil conditions favorable for weed growth.
- ☆ Informed decision-making on soil amendments to suppress weeds naturally.

- **Example**: Nano sensors can detect nitrogen or phosphorus levels in the soil, enabling farmers to adjust fertilization practices to discourage weed growth.

4. Improved Crop Health and Weed Stress Detection

Nano sensors can monitor crop health by detecting stress signals caused by weed competition. These sensors measure parameters such as chlorophyll content, photosynthetic efficiency, and water uptake, providing insights into how weeds are affecting crop performance.

- **Key Insight**: By understanding the impact of weeds on crop health, farmers can prioritize weed control measures in areas where crops are most stressed.
- **Example**: Nano sensors integrated into plant leaves can detect changes in photosynthesis rates due to shading by weeds.

5. Environmental Sustainability

The use of nano sensors in precision agriculture promotes sustainable farming practices by reducing chemical inputs and minimizing their impact on ecosystems.

Advantages

- Lower herbicide runoff into water bodies.
- Reduced soil degradation caused by excessive chemical use.
- Preservation of beneficial organisms in the soil and surrounding environment.
- **Scientific Principle**: Nano sensors operate at a molecular level, ensuring precise detection and targeted action, which reduces the ecological footprint of agricultural practices.

6. Data-Driven Decision Making

Nano sensors generate vast amounts of data that can be integrated into decision-support systems and farm management software. This data-driven approach enables farmers to make informed decisions about weed control strategies.

Applications

- Predictive modeling for weed outbreaks based on environmental data.
- Optimization of input use, such as herbicides and fertilizers.
- **Example**: Internet of Things (IoT) platforms connected to nano sensors provide real-time alerts and actionable insights for weed management.

7. Cost-Effectiveness

Although the initial investment in nano sensor technology may be high, the long-term cost savings are substantial. By reducing input costs, labor, and crop losses due to weeds, nano sensors offer a high return on investment.

Economic Benefits

- Reduced herbicide expenditure through targeted application.
- Increased crop yields due to better weed control and crop health.
- **Example**: Farmers using nano-enabled precision agriculture systems report up to 30 per cent savings in herbicide costs (Kumar *et al.*, 2021).

8. Integration with Advanced Technologies

Nano sensors seamlessly integrate with other advanced agricultural technologies, such as drones, robotics, and artificial intelligence (AI). This synergy enhances the efficiency and accuracy of weed management practices.

- **Example**: AI-powered drones equipped with nano sensors can autonomously identify and treat weed-infested areas, reducing the need for manual labor.
- **Future Potential**: The integration of nano sensors with blockchain technology could enable traceability in weed management practices, ensuring transparency and compliance with regulatory standards.

F. Nano Sensors for Real-Time Weed Detection

The integration of nanotechnology into agriculture has revolutionized weed management by enabling precise, real-time weed detection. Nano sensors, which operate at the nanoscale (1–100 nm), offer unparalleled sensitivity and specificity, making them a promising tool for modern weed control. These advanced devices can detect biochemical and physiological differences between crops and weeds, allowing for targeted interventions that minimize herbicide use and environmental impact.

Principles of Nano Sensors in Weed Detection

Nano sensors function by detecting specific signals or biomarkers associated with weeds. These signals can include volatile organic compounds (VOCs), chlorophyll fluorescence, or other unique biochemical signatures. The working mechanism typically involves:

- **Recognition Element**: A component (*e.g.*, nanomaterials, enzymes, or antibodies) that interacts with the target molecule or signal.
- **Transduction Mechanism**: Converts the interaction into a measurable signal, such as an electrical, optical, or chemical response.

☆ **Signal Processing Unit**: Analyzes and interprets the data for real-time decision-making.

For example, carbon nanotubes and gold nanoparticles are commonly used in nano sensors due to their high surface area, conductivity, and ability to bind with specific molecules.

Applications in Weed Management

Nano sensors have several applications in real-time weed detection, including:

1. **Discrimination Between Crops and Weeds:** Nano sensors can identify weeds based on their unique VOC emissions or spectral properties. For instance, weeds often emit distinct VOCs during their growth stages, which can be detected using nano sensors integrated into drones or robotic platforms.
2. **Monitoring Herbicide Resistance:** Nano sensors can detect molecular markers of herbicide resistance in weeds, such as mutations in the acetolactate synthase (ALS) or EPSPS genes. This allows for early identification of resistant weed populations and informed herbicide selection.
3. **Precision Herbicide Application:** By integrating nano sensors with GPS-enabled sprayers, herbicides can be applied only to weed-infested areas, reducing chemical usage and preventing crop damage.
4. **Soil and Environmental Monitoring:** Nano sensors can also detect weed seeds or seedlings in the soil by sensing specific metabolites or enzymes. This enables pre-emergence weed control strategies.

Advantages of Nano Sensors

☆ **High Sensitivity**: Nano sensors can detect minute concentrations of weed-specific biomarkers.

☆ **Real-Time Monitoring**: Enables immediate detection and response, reducing weed competition with crops.

☆ **Environmental Sustainability**: Minimizes herbicide overuse, protecting soil health and biodiversity.

☆ **Cost-Effectiveness**: Reduces labor and input costs by automating weed detection and management.

Challenges and Future Prospects

Despite their potential, nano sensors face challenges such as high production costs, scalability issues, and the need for robust field validation. However, advancements in nanofabrication and material science are expected to address these limitations. Future research may focus on integrating nano sensors with artificial intelligence (AI) and machine learning (ML) for enhanced weed detection accuracy.

Case Studies and Examples

- ☆ **Graphene-Based Nano Sensors**: Researchers have developed graphene-based sensors capable of detecting VOCs emitted by weeds like *Amaranthus palmeri* (Palmer amaranth) with high accuracy (Chauhan *et al.*, 2021).
- ☆ **Nano Sensors in UAVs**: Unmanned aerial vehicles (UAVs) equipped with nano sensors have been successfully deployed to map weed infestations in large agricultural fields (Singh *et al.*, 2022).

G. What are Nano Herbicides?

Nano herbicides represent an innovative advancement in weed management, leveraging nanotechnology to enhance the efficacy, precision, and environmental safety of herbicidal applications. These formulations utilize nanoparticles or nanostructures to deliver active ingredients in a controlled and targeted manner, addressing several limitations of conventional herbicides.

Key Features of Nano Herbicides

1. **Enhanced Solubility and Bioavailability:** Many herbicides have low water solubility, which limits their effectiveness. Nano herbicides improve solubility by encapsulating the active ingredient in nanocarriers such as liposomes, micelles, or polymeric nanoparticles. This increases bioavailability, ensuring better absorption by the target weed species.
2. **Controlled Release Mechanism:** Nano herbicides are designed to release their active ingredients gradually over time. This controlled release reduces the frequency of application, minimizes herbicide wastage, and ensures prolonged weed control. For example, silica nanoparticles have been used to encapsulate herbicides, allowing for sustained release under field conditions.
3. **Targeted Delivery:** Nanotechnology enables the development of herbicides that specifically target weeds while sparing crops and non-target organisms. Functionalized nanoparticles, such as those coated with ligands or antibodies, can bind selectively to weed-specific receptors. This reduces off-target effects and environmental contamination.
4. **Reduced Environmental Impact:** Conventional herbicides often lead to runoff, leaching, and accumulation in soil and water bodies, causing environmental harm. Nano herbicides, due to their precision and controlled release, minimize these risks. For instance, studies have

shown that nano-encapsulated atrazine reduces leaching compared to its conventional formulation (Kah *et al.*, 2018).

5. **Improved Stability:** Nano formulations protect active ingredients from degradation caused by environmental factors such as UV radiation, temperature fluctuations, and microbial activity. This ensures the herbicide remains effective for longer periods.

Examples of Nano Herbicides

- **Nano-encapsulated Glyphosate**: Encapsulation in polymeric nanoparticles enhances glyphosate's stability and reduces its environmental footprint.
- **Chitosan-based Nano Herbicides**: Chitosan nanoparticles have been used to deliver herbicides with improved adhesion to plant surfaces and reduced toxicity to non-target organisms.

Challenges and Future Prospects

While nano herbicides offer significant advantages, challenges such as high production costs, regulatory hurdles, and potential nanoparticle toxicity to soil microbiota remain. Ongoing research aims to address these issues and optimize formulations for large-scale agricultural use.

H. Mechanism of Action of Nano Herbicides

Nano herbicides represent a cutting-edge advancement in modern weed management, leveraging nanotechnology to enhance herbicide efficacy, reduce environmental impact, and improve weed control precision. These formulations utilize nanoscale carriers or active ingredients to deliver herbicidal compounds more effectively to target weeds. Understanding their mechanism of action is crucial for optimizing their application and ensuring sustainable weed management practices.

Key Components of Nano Herbicides

Nano herbicides typically consist of the following components:

- **Nanocarriers**: Materials such as liposomes, polymeric nanoparticles, silica nanoparticles, or carbon nanotubes that encapsulate or bind herbicide molecules.
- **Active Ingredients (AIs)**: Herbicidal compounds that disrupt specific physiological or biochemical processes in weeds.
- **Surface Modifications**: Functional groups or coatings that enhance the stability, solubility, and target specificity of the nanoformulation.

Mechanisms of Action

The mechanism of action of nano herbicides can be broadly divided into the following stages:

1. Enhanced Delivery and Uptake

Nano herbicides improve the delivery of active ingredients to target weeds by overcoming barriers such as poor solubility, rapid degradation, or limited absorption.

- ☆ **Improved Solubility**: Nanocarriers increase the solubility of hydrophobic herbicides, ensuring better dispersion in aqueous environments.
- ☆ **Leaf Penetration**: Due to their small size (typically <100 nm), nanoparticles can penetrate plant cuticles and cell walls more effectively than conventional formulations. This enhances the uptake of active ingredients into weed tissues.
- ☆ **Controlled Release**: Nano herbicides often exhibit controlled or sustained release properties, ensuring a steady supply of the active ingredient over time. This reduces the frequency of application and minimizes environmental losses.

2. Targeted Action

Nano herbicides can be engineered for site-specific action, reducing off-target effects and improving weed control efficiency.

- ☆ **Selective Binding**: Surface modifications enable nanocarriers to bind specifically to receptors or biomolecules unique to weed species. For example, functionalized nanoparticles can target specific enzymes involved in photosynthesis or amino acid synthesis.
- ☆ **pH-Responsive Release**: Some nano herbicides are designed to release their active ingredients in response to specific pH conditions, such as those found in weed tissues.

3. Enhanced Herbicidal Activity

The nanoscale delivery of active ingredients enhances their bioavailability and interaction with target sites, leading to improved herbicidal activity.

- ☆ **Increased Cellular Uptake**: Nanoparticles facilitate the intracellular transport of active ingredients, ensuring their delivery to critical organelles such as chloroplasts or mitochondria.
- ☆ **Synergistic Effects**: Nanocarriers can be co-loaded with multiple active ingredients or adjuvants, resulting in synergistic effects that enhance weed control efficacy.

4. Reduced Environmental Impact

Nano herbicides minimize environmental contamination and non-target effects through their precision delivery mechanisms.

- **Lower Doses**: The improved efficiency of nano herbicides allows for lower application rates compared to conventional formulations.
- **Reduced Drift and Runoff**: Encapsulation of active ingredients reduces the risk of herbicide drift or leaching into non-target areas.

Examples of Nano Herbicides

- **Atrazine-Loaded Nanoparticles**: Studies have shown that polymeric nanoparticles loaded with atrazine exhibit enhanced weed control efficacy and reduced environmental persistence (Kah *et al.*, 2013).
- **Glyphosate Nanoformulations**: Nano glyphosate formulations demonstrate improved uptake and translocation in weeds, leading to higher efficacy at reduced doses (De Oliveira *et al.*, 2021).

Scientific Principles Underlying Nano Herbicides

- **Nanoparticle-Plant Interactions**: The small size and high surface area of nanoparticles facilitate their interaction with plant tissues, enhancing herbicide delivery.
- **Controlled Release Kinetics**: The release of active ingredients from nanocarriers follows specific kinetic models, such as diffusion or degradation-controlled release, ensuring sustained action.
- **Surface Chemistry**: Functionalization of nanoparticles with hydrophilic or hydrophobic groups influences their solubility, stability, and target specificity.

Future Potential

The integration of nanotechnology with herbicide development holds immense potential for addressing challenges such as herbicide resistance, environmental contamination, and the need for sustainable agriculture. Continued research and innovation are essential to fully realize the benefits of nano herbicides in modern weed management.

I. Advantages of Nano Herbicides in Weed Management

The advent of nanotechnology has revolutionized various sectors, including agriculture. Nano herbicides, a novel application of nanotechnology, have emerged as a promising solution for effective weed management. These herbicides utilize nanoparticles to enhance the delivery, efficacy, and environmental safety of active ingredients. Below are the key advantages of nano herbicides in weed management, supported by scientific principles and practical examples.

1. Enhanced Herbicide Efficiency

- **Improved Solubility and Bioavailability**: Nano herbicides improve the solubility of poorly water-soluble active ingredients, ensuring better absorption by target weeds. For instance, encapsulating herbicides in nanocarriers like liposomes or polymeric nanoparticles enhances their bioavailability, leading to higher efficacy at lower doses.
- **Targeted Delivery**: Nanoparticles can be engineered to deliver herbicides specifically to weed tissues, minimizing off-target effects. For example, magnetic nanoparticles can be guided using external magnetic fields to deposit herbicides precisely on weed-infested areas (Kah *et al.*, 2013).
- **Controlled Release**: Nano herbicides allow for controlled and sustained release of active ingredients over time. This reduces the need for repeated applications, saving costs and labor while maintaining long-term weed suppression.

2. Reduced Environmental Impact

- **Minimized Chemical Runoff**: Conventional herbicides often lead to significant chemical runoff, contaminating water bodies and harming aquatic ecosystems. Nano herbicides, with their controlled release properties, reduce the likelihood of leaching and runoff, thus safeguarding the environment.
- **Lower Application Rates**: Due to their enhanced efficiency, nano herbicides require smaller quantities of active ingredients compared to traditional formulations. This reduces the overall chemical load on the environment.
- **Biodegradable Nanocarriers**: Many nano herbicides use biodegradable materials such as chitosan or lignin as carriers, which decompose naturally without leaving harmful residues in the soil.

3. Resistance Management

- **Overcoming Herbicide Resistance**: The precise delivery and controlled release of nano herbicides can help manage herbicide-resistant weed populations. By ensuring consistent exposure of weeds to the active ingredient, nano herbicides reduce the chances of resistance development.
- **Combination Formulations**: Nanotechnology enables the co-delivery of multiple active ingredients in a single formulation. This multi-target approach can effectively combat resistant weed species by attacking different physiological pathways simultaneously.

4. Improved Safety for Non-Target Organisms

- **Selective Action**: Nano herbicides can be designed to release active ingredients only in specific environmental conditions, such as pH or temperature changes, ensuring minimal impact on non-target plants and beneficial organisms.
- **Reduced Drift**: The use of nano-sized particles reduces the likelihood of herbicide drift during application, protecting nearby crops, pollinators, and other non-target species.

5. Economic Benefits

- **Cost-Effectiveness**: Although the initial production cost of nano herbicides may be higher, their long-term benefits, such as reduced application rates, fewer reapplications, and lower environmental remediation costs, make them economically viable.
- **Extended Shelf Life**: Nano formulations often exhibit improved stability, extending the shelf life of herbicides and reducing wastage.

6. Compatibility with Precision Agriculture

- Nano herbicides align seamlessly with precision agriculture technologies. For instance, drones equipped with sensors can identify weed-infested areas, and nano herbicides can be applied in precise quantities to those specific zones. This targeted approach minimizes wastage and maximizes efficiency.

7. Potential for Sustainable Agriculture

- **Integration with Organic Farming**: Nano herbicides using natural carriers and bio-based active ingredients can be integrated into organic farming systems, promoting sustainable agriculture.
- **Reduction in Carbon Footprint**: By decreasing the frequency of herbicide applications and the amount of chemical usage, nano herbicides contribute to a lower carbon footprint in weed management practices.

Examples of Nano Herbicides in Practice

- **Nano-Encapsulated Glyphosate**: Studies have shown that nano-encapsulation of glyphosate improves its efficacy against resistant weeds while reducing environmental contamination (Grillo *et al.*, 2014).
- **Chitosan-Based Nano Herbicides**: Chitosan nanoparticles have been used to deliver herbicides with enhanced stability and biodegradability, making them an eco-friendly alternative.

J. Environmental Impact of Nano Herbicides

The advent of nanotechnology in agriculture has led to the development of nano herbicides, which promise precision weed control with minimal environmental impact. Nano herbicides are herbicidal formulations that utilize nanoparticles as carriers or active ingredients to enhance the efficacy, reduce dosage, and mitigate the environmental footprint of traditional herbicides. However, their environmental implications must be critically assessed to ensure their sustainability and safe adoption in modern weed management practices.

Advantages of Nano Herbicides in Reducing Environmental Impact

Nano herbicides offer several advantages over conventional herbicides, particularly in terms of their environmental footprint:

- ☆ **Target-Specific Action**: Nano herbicides can be engineered for controlled release and site-specific delivery, reducing off-target effects and minimizing damage to non-target plants, soil microbes, and beneficial organisms. For instance, encapsulating herbicides in nanocarriers like liposomes or polymeric nanoparticles ensures their release only in the presence of specific environmental triggers, such as pH or enzymatic activity in target weeds (Kah *et al.*, 2018).
- ☆ **Reduced Dosage Requirements**: Due to their enhanced bioavailability and precision, nano herbicides often require lower application rates compared to conventional formulations. This reduces the overall chemical load on the environment and mitigates risks of soil and water contamination.
- ☆ **Minimized Volatilization and Leaching**: Nano herbicides exhibit reduced volatilization and leaching due to their controlled-release properties. This helps in preventing herbicide drift, groundwater contamination, and the accumulation of toxic residues in the ecosystem.

Potential Risks and Concerns

Despite their potential benefits, nano herbicides pose certain environmental risks that need careful evaluation:

1. **Nanoparticle Persistence**: The long-term persistence of nanoparticles in soil and water ecosystems is a major concern. Some nanoparticles, such as metal oxides or carbon-based nanomaterials, may accumulate in the environment, potentially disrupting microbial communities and soil health.
2. **Toxicity to Non-Target Organisms**: While nano herbicides are designed for target-specific action, unintended exposure to non-target organisms, including soil fauna, aquatic species, and pollinators, remains a possibility. Studies have shown that certain nanoparticles, such as silver or zinc oxide, can exhibit toxicity to beneficial soil

microbes and aquatic organisms at higher concentrations (Rico *et al.*, 2011).

3. **Bioaccumulation and Biomagnification**: The small size and high reactivity of nanoparticles raise concerns about their potential to bioaccumulate in food chains. For example, nanoparticles that enter aquatic systems may be ingested by plankton and subsequently magnified through trophic levels, posing risks to higher organisms, including humans.
4. **Unintended Interactions with Soil Chemistry**: Nanoparticles can interact with soil components, altering nutrient availability, pH, or microbial activity. These changes may have cascading effects on soil fertility and crop productivity.

Regulatory and Monitoring Challenges

The unique properties of nano herbicides necessitate the development of specialized regulatory frameworks and monitoring protocols:

- **Lack of Standardized Testing**: Current environmental risk assessment protocols for agrochemicals may not adequately address the unique behavior of nanoparticles. Parameters such as particle size, shape, surface charge, and aggregation must be considered to accurately predict their environmental fate.
- **Challenges in Detection and Quantification**: The detection and quantification of nanoparticles in complex environmental matrices, such as soil and water, remain technically challenging. Advanced analytical techniques, such as transmission electron microscopy (TEM) and dynamic light scattering (DLS), are required but are often expensive and time-consuming.
- **Need for Lifecycle Analysis**: Comprehensive lifecycle assessments are essential to evaluate the environmental impact of nano herbicides from production to degradation. This includes assessing the energy and resource requirements for nanoparticle synthesis as well as their end-of-life behavior.

Future Directions for Sustainable Use

To ensure the sustainable use of nano herbicides, the following measures should be prioritized:

- **Development of Biodegradable Nanocarriers**: Research should focus on designing nanocarriers that degrade into non-toxic byproducts, minimizing their environmental persistence.
- **Integrated Weed Management (IWM) Approaches**: Nano herbicides should be incorporated into broader IWM strategies that combine

cultural, mechanical, and biological weed control methods to reduce reliance on chemical inputs.

- **Stakeholder Collaboration**: Collaboration between researchers, policymakers, and industry stakeholders is essential to establish robust safety guidelines and promote the responsible use of nano herbicides.

K. Challenges in Nano Sensor and Nano Herbicide Implementation

The advent of nanotechnology in agriculture has opened new frontiers for precision weed management. Nano sensors and nano herbicides offer promising solutions for detecting and controlling weeds with unprecedented accuracy and efficiency. However, their implementation is fraught with several challenges that hinder large-scale adoption. This section explores these challenges in detail, focusing on technical, economic, environmental, and regulatory aspects.

1. Technical Challenges

a. Nano Sensor Sensitivity and Specificity

- Nano sensors designed for weed detection must exhibit high sensitivity to differentiate between crops and weeds effectively. However, achieving this level of specificity is challenging due to the complex biochemical and morphological similarities between certain weed species and crops.
- Environmental factors such as soil type, moisture levels, and temperature fluctuations can interfere with the performance of nano sensors, leading to false positives or negatives.

b. Nano Herbicide Delivery Mechanisms

- Nano herbicides rely on advanced delivery systems, such as encapsulation or controlled-release mechanisms, to target weeds precisely. However, ensuring uniform distribution and effective uptake by weed tissues remains a significant hurdle.
- The interaction of nano herbicides with soil particles and microbial communities can alter their efficacy, requiring further research to optimize formulations.

c. Scalability of Production

- The production of nano sensors and nano herbicides at a commercial scale poses technical difficulties. Current manufacturing techniques, such as chemical vapor deposition or sol-gel processes, are often expensive and time-consuming, limiting scalability.

2. Economic Challenges

a. High Initial Costs

- The development and deployment of nano-based technologies involve substantial initial investments in research, infrastructure, and equipment. This cost is often prohibitive for small and medium-scale farmers.
- The lack of economies of scale in the early stages of adoption further inflates the cost of nano sensors and nano herbicides.

b. Limited Market Penetration

- The agricultural sector, particularly in developing countries, is characterized by low technological adoption rates. Farmers may be reluctant to invest in nano-based solutions due to a lack of awareness or perceived cost-benefit imbalances.

3. Environmental Challenges

a. Potential Toxicity of Nanomaterials

- The environmental impact of nanomaterials used in sensors and herbicides is not yet fully understood. Studies have raised concerns about the potential toxicity of nanoparticles to non-target organisms, including beneficial soil microbes, insects, and aquatic ecosystems (Kah *et al.*, 2018).
- The persistence of nanoparticles in the environment could lead to bioaccumulation and unforeseen ecological consequences.

b. Soil and Water Contamination

- Improper application or overuse of nano herbicides may result in the contamination of soil and water resources. This could have long-term implications for soil health and water quality, necessitating stringent application protocols.

4. Regulatory and Ethical Challenges

a. Lack of Standardized Guidelines

- The regulatory framework for nano-based agricultural products is still in its infancy. There is a lack of standardized guidelines for testing, approval, and labeling of nano sensors and nano herbicides, creating uncertainty for manufacturers and users alike.
- Regulatory agencies face difficulties in assessing the safety and efficacy of these products due to the unique properties of nanomaterials.

b. Ethical Concerns

- ✰ The use of nanotechnology in agriculture raises ethical questions about its potential impact on traditional farming practices and rural livelihoods. There is also a risk of monopolization by large agribusinesses, which could marginalize small-scale farmers.

5. Knowledge and Skill Gaps

- ✰ The successful implementation of nano sensors and nano herbicides requires a workforce with specialized knowledge in nanotechnology, agronomy, and environmental science. However, there is a significant gap in training and education programs tailored to these emerging technologies.
- ✰ Extension services and farmer training programs are often inadequate, further limiting the adoption of nano-based weed management solutions.

6. Public Perception and Acceptance

- ✰ Public skepticism about the safety and ethical implications of nanotechnology can hinder its acceptance in agriculture. Misinformation and lack of awareness about the benefits of nano sensors and nano herbicides exacerbate this issue.
- ✰ Building trust through transparent communication and stakeholder engagement is crucial for overcoming these barriers.

L. Future Trends in Nano Technology for Weed Control

Nanotechnology has emerged as a transformative tool in agriculture, offering innovative solutions for sustainable weed management. By leveraging nanoscale materials and systems, researchers are exploring precise, efficient, and environmentally friendly alternatives to conventional herbicides. Below, we discuss the latest advancements and future trends in nanotechnology for weed control.

1. Nano-Herbicides for Targeted Weed Control

Nano-herbicides are being developed to enhance the efficacy of active ingredients while minimizing their environmental impact. These formulations use nanoparticles as carriers to deliver herbicides directly to the target site, reducing off-target effects and lowering the required dosage. Key advancements include:

- ✰ **Controlled Release Systems**: Nanoparticles can be engineered to release herbicides in response to specific environmental triggers, such as pH, temperature, or light. This ensures precise application and prolonged activity.

- **Improved Solubility**: Nanotechnology enhances the solubility and stability of poorly water-soluble herbicides, increasing their bioavailability and effectiveness.

For example, silica nanoparticles have been used to encapsulate herbicides like atrazine, enabling slow and controlled release, thereby reducing leaching and runoff (Kah *et al.*, 2018).

2. Nano-Sensors for Early Weed Detection

Nano-sensors are revolutionizing weed management by enabling real-time monitoring and early detection of weed infestations. These sensors can:

- Detect specific weed biomarkers or stress signals emitted by crops due to weed competition.
- Provide precise data for site-specific herbicide application, reducing chemical usage and labor costs.

For instance, carbon-based nanomaterials, such as graphene oxide, are being explored for their high sensitivity in detecting biochemical changes in plants under stress.

3. Bio-Nanotechnology for Eco-Friendly Solutions

The integration of nanotechnology with biological systems is paving the way for bio-nanotechnology in weed control. This approach focuses on:

- **Nano-Bioherbicides**: Using nanoparticles to deliver natural herbicidal compounds, such as allelopathic extracts, for sustainable weed suppression.
- **Biodegradable Nanomaterials**: Developing eco-friendly nanoparticles that degrade into non-toxic byproducts, reducing environmental contamination.

4. Challenges and Future Directions

While nanotechnology holds immense potential, challenges such as regulatory approval, cost of production, and potential toxicity to non-target organisms need to be addressed. Future research is expected to focus on:

- Developing multifunctional nanoparticles that combine weed detection, herbicide delivery, and environmental monitoring.
- Scaling up production processes to make nano-herbicides economically viable for widespread adoption.

Chapter 10

Biotechnological Innovations

Biotechnology has revolutionized weed management by introducing precise, sustainable, and innovative solutions to combat weed infestations. These advancements integrate genetic engineering, molecular biology, and bioinformatics to enhance crop resistance and target weed control mechanisms effectively.

1. Herbicide-Tolerant Crops (HTCs)

- ✰ Herbicide-tolerant crops, such as glyphosate-resistant soybean and maize, allow farmers to apply non-selective herbicides without damaging the crop. This reduces the need for multiple herbicide applications and minimizes labor costs.
- ✰ Advances in CRISPR-Cas9 gene-editing technology have enabled the development of more precise HTCs, reducing off-target effects and improving crop safety (Schmidt *et al.,* 2020).

2. RNA Interference (RNAi) Technology

- ✰ RNAi-based weed management involves silencing essential genes in weeds, leading to their growth inhibition or death. This method is species-specific, reducing the risk of harming non-target organisms.
- ✰ For example, research has demonstrated the potential of RNAi to control problematic weeds like Palmer amaranth (Amaranthus palmeri) by targeting genes critical for its survival (Li *et al.,* 2021).

3. Bioherbicides

- ✰ Biotechnological advancements have improved the development of bioherbicides derived from natural pathogens or microbial metabolites.

These eco-friendly alternatives, such as Mycoherbicides, target specific weeds without affecting crops or the environment.

- ☆ For instance, *Phytophthora palmivora* has been successfully used to manage milkweed vine in tropical regions.

4. Weed Genomics and Bioinformatics

- ☆ Genomic sequencing of weed species provides insights into their resistance mechanisms, enabling the design of targeted control strategies.
- ☆ Bioinformatics tools, such as genome-wide association studies (GWAS), help identify resistance genes in weeds, facilitating the development of novel herbicides.

Biotechnological innovations are paving the way for sustainable and precise weed management practices, addressing challenges like herbicide resistance and environmental safety.

A. Genetically Modified (GM) Crops

Genetically Modified (GM) crops have emerged as a transformative tool in modern agriculture, offering innovative solutions to challenges such as weed management, pest resistance, and environmental sustainability. By introducing specific genetic traits into crops, scientists have been able to enhance their resistance to herbicides, improve crop yields, and reduce the need for chemical inputs. This section explores the role of GM crops in weed management, their mechanisms, advantages, limitations, and their implications for sustainable agriculture.

Introduction to GM Crops

Genetically Modified (GM) crops are plants whose genetic material has been altered using modern biotechnology techniques, such as recombinant DNA technology. These modifications enable the introduction of desirable traits, such as herbicide tolerance or pest resistance, which are not achievable through traditional breeding methods. The introduction of GM crops in weed science is a new concept which is gaining popularity at a very fast rate.

The first genetically modified crop to be grown commercially was the Flavr Savr tomato, introduced by the U.S. company Calgene in 1994. This tomato was engineered to ripen more slowly, thereby extending its shelf life. It opened the door for subsequent GM crops like Bt corn, Roundup Ready soybeans, and others that followed in later years. Now this concept is used in many field crops.

Table 10.1 shows the rate of global adoption of each crop for the year 2023. Cotton contains the highest rate of GM per cent utilisation at 76.0 per cent of the world's total cotton area, followed by soybean at 72.4 per cent, maize at 34.0

per cent and canola at 24.0 per cent. Following this, the rate of global adoption falls away quickly.

Table 10.1. Global Adoption of GM Crops

Crop	GM Area (Ha m.)	Total Area (Ha m.)	GM % Share
Cotton	24.1	31.7	76.0
Soybean	100.9	139.4	72.4
Maize	69.3	203.5	34.0
Canola	10.2	42.5	24.0
Sugar beet	0.5	4.3	11.0
Alfalfa	1.2	35.0	3.4
Sugarcane	0.06	26.1	0.2
Brinjal	0.003	1.9	0.2
Rice	0.05	165.1	0.03
Wheat	0.04	222.7	0.02
Total	**206.26**	**872.16**	**23.7**

Key GM Traits for Weed Management

- ✰ **Herbicide Tolerance (HT):** Enables crops to withstand specific herbicides, allowing farmers to control weeds without damaging the crop.
- ✰ **Stacked Traits:** Combines herbicide tolerance with other traits, such as insect resistance, for broader utility.

Examples of widely adopted GM crops include herbicide-tolerant soybean, maize, cotton, and canola. These crops have revolutionized weed control practices by simplifying weed management and reducing labor-intensive practices like mechanical weeding.

Mechanisms of Herbicide-Tolerant GM Crops

Herbicide-tolerant GM crops are engineered to survive the application of specific herbicides, which would otherwise harm or kill the crop. The two most common herbicide-tolerant systems are:

1. Glyphosate-Tolerant Crops

- ✰ Glyphosate is a broad-spectrum herbicide that inhibits the enzyme 5-enolpyruvylshikimate-3-phosphate synthase (EPSPS), essential for amino acid synthesis in plants.

- GM crops, such as Roundup Ready® soybean, are engineered to express a modified EPSPS enzyme, making them resistant to glyphosate application. This allows farmers to apply glyphosate post-emergence to control a wide range of weeds without affecting the crop.

2. Glufosinate-Tolerant Crops

- Glufosinate inhibits glutamine synthetase, an enzyme critical for nitrogen metabolism in plants.
- GM crops, such as LibertyLink® maize, are engineered to detoxify glufosinate, enabling effective weed control.

These mechanisms provide farmers with flexibility in weed management, allowing for the use of non-selective herbicides that target a broad spectrum of weeds.

Advantages of GM Crops in Weed Management

The adoption of GM crops with herbicide tolerance has provided several advantages for weed control, including:

Simplified Weed Control

- Farmers can use a single herbicide to manage a wide range of weed species, reducing the complexity of weed management programs.
- Post-emergence herbicide applications are possible, allowing for better timing and efficiency.

Reduced Tillage Requirements

- Herbicide-tolerant crops support no-till or reduced-till farming systems, which help conserve soil structure, reduce erosion, and improve soil health.

Cost-Effectiveness

- Reduced reliance on multiple herbicides and mechanical weeding lowers production costs.
- Labor savings are significant, especially in large-scale farming operations.

Improved Crop Yields

- Effective weed control minimizes competition for nutrients, water, and light, leading to higher crop productivity.

Environmental Benefits

- Reduced tillage practices associated with GM crops decrease greenhouse gas emissions and fuel consumption.

- ☆ Targeted herbicide use minimizes the overall chemical load on the environment.

Challenges and Limitations

Despite their benefits, GM crops also present challenges and limitations that must be addressed for sustainable weed management:

1. Herbicide Resistance in Weeds

- ☆ Over-reliance on a single herbicide, such as glyphosate, has led to the evolution of herbicide-resistant weed populations. For example, glyphosate-resistant Palmer amaranth (*Amaranthus palmeri*) has become a major concern in the United States (Heap, 2023).
- ☆ Integrated Weed Management (IWM) strategies, including crop rotation and the use of multiple herbicide modes of action, are essential to mitigate resistance.

2. Gene Flow and Non-Target Effects

- ☆ Cross-pollination between GM crops and related wild species can result in the unintended spread of herbicide-tolerant traits, potentially creating "superweeds."
- ☆ Non-target effects, such as impacts on soil microbiota and biodiversity, require further study.

3. Regulatory and Public Perception Issues

- ☆ Regulatory frameworks for GM crops vary globally, affecting their adoption and market access.
- ☆ Public concerns about the safety and environmental impact of GM crops continue to influence their acceptance.

4. Cost of Development and Access

- ☆ The development of GM crops is expensive, and intellectual property rights can limit access for smallholder farmers in developing countries.

Best Practices for Using GM Crops in Weed Management

To maximize the benefits of GM crops while minimizing risks, the following best practices should be adopted:

Rotate Herbicide Modes of Action

- ☆ Avoid repeated use of the same herbicide to delay the development of resistant weed populations.
- ☆ Use pre-emergence herbicides in combination with post-emergence applications.

Adopt Integrated Weed Management (IWM)

- ☆ Combine cultural, mechanical, and chemical weed control methods to reduce reliance on herbicides.
- ☆ Use crop rotation and cover crops to suppress weed growth.

Monitor and Manage Resistance

- ☆ Regularly scout fields for signs of herbicide-resistant weeds and implement resistance management strategies.
- ☆ Use herbicide mixtures or sequential applications with different modes of action.

Promote Stewardship Programs

- ☆ Educate farmers on the responsible use of GM crops and herbicides.
- ☆ Encourage adherence to refuge requirements in areas where stacked traits are used to manage pest resistance.

Examples of GM Crops for Weed Management

1. Roundup Ready® Soybean

- ☆ Developed by Monsanto, this glyphosate-tolerant soybean variety has been widely adopted in countries like the United States, Brazil, and Argentina.
- ☆ It allows for effective post-emergence weed control, reducing the need for pre-emergence herbicides.

2. LibertyLink® Maize

- ☆ This glufosinate-tolerant maize variety provides an alternative to glyphosate-based systems, offering flexibility in herbicide choice.
- ☆ It is particularly useful in regions where glyphosate-resistant weeds are prevalent.

3. XtendFlex® Cotton

- ☆ Combines tolerance to glyphosate, glufosinate, and dicamba, providing multiple options for weed control.
- ☆ This stacked trait technology helps manage herbicide-resistant weeds more effectively.

Future Prospects

The future of GM crops in weed management lies in the development of new traits and technologies to address emerging challenges. Some promising areas of research include:

- ☆ **Next-Generation Herbicide Tolerance**: Development of crops tolerant to new herbicide chemistries with novel modes of action to combat resistant weeds.
- ☆ **Gene Editing Technologies**: Tools like CRISPR-Cas9 offer precise and cost-effective methods for introducing desirable traits into crops.
- ☆ **Biotechnology for Weed Suppression**: Engineering crops to produce allelopathic compounds that suppress weed growth naturally.
- ☆ **Sustainable GM Solutions**: Integration of GM crops with ecological approaches, such as precision agriculture and digital weed mapping, to optimize weed control while minimizing environmental impact.

B. Bioherbicides

Bioherbicides represent an innovative and environmentally sustainable approach to weed management. Derived from natural organisms such as fungi, bacteria, or plant-derived compounds, bioherbicides offer a targeted mechanism for controlling weeds while minimizing the environmental and health risks associated with synthetic chemical herbicides. This section explores the principles, mechanisms, examples, and applications of bioherbicides in modern weed management.

Principles of Bioherbicides

Bioherbicides are based on the use of living organisms or their metabolites to suppress weed growth. Unlike synthetic herbicides, which often have broad-spectrum activity, bioherbicides are typically species-specific, targeting particular weeds without harming non-target plants or ecosystems. The principles underlying bioherbicide action include:

- ☆ **Pathogenicity:** Many bioherbicides rely on pathogens, such as fungi or bacteria, that infect and kill specific weed species.
- ☆ **Allelopathy:** Some bioherbicides utilize plant-derived chemicals that inhibit the germination or growth of weeds.
- ☆ **Host Specificity:** Bioherbicides are designed to act selectively on specific weed species, reducing collateral damage to crops and beneficial plants.

The development of bioherbicides often involves isolating and identifying natural organisms or compounds with herbicidal properties, followed by formulation and application strategies to maximize efficacy.

Mechanisms of Action

Bioherbicides work through diverse mechanisms, depending on the organism or compound used. The primary modes of action include:

1. **Infection and Colonization:** Fungal bioherbicides, such as *Colletotrichum gloeosporioides*, infect weed tissues, leading to necrosis and plant death. These pathogens often produce enzymes and toxins that degrade plant cell walls and disrupt physiological processes.
2. **Production of Phytotoxins:** Certain bacteria, such as *Pseudomonas fluorescens*, release phytotoxic compounds that interfere with photosynthesis, respiration, or other metabolic pathways in weeds.
3. **Allelopathic Interference:** Plant-derived bioherbicides, such as those containing allelopathic compounds like juglone (from walnut trees) or sorgoleone (from sorghum), inhibit seed germination or root elongation in competing weeds.
4. **Competition for Resources:** Some bioherbicides act by outcompeting weeds for essential resources such as nutrients, water, or light, thereby suppressing their growth.

Examples of Bioherbicides

Several bioherbicides have been commercialized or are under development for weed control. Below are notable examples:

1. **Devine® (*Phytophthora palmivora*):** A fungal bioherbicide used to control *Striga* spp. (witchweed) in tropical and subtropical regions. It infects the roots of the parasitic weed, reducing its ability to establish and reproduce.
2. **Collego® (*Colletotrichum gloeosporioides*):** A fungal bioherbicide targeting *Aeschynomene virginica* (northern jointvetch), a problematic weed in rice and soybean fields.
3. **BioMal® (*Colletotrichum coccodes*):** A fungal bioherbicide developed for the control of *Malva pusilla* (round-leaved mallow) in agricultural systems.
4. **Sorgoleone:** A natural allelopathic compound produced by sorghum roots, which has shown potential as a bioherbicide for controlling grassy weeds.
5. **Dr. Biosedge:** A bacterial bioherbicide containing *Pseudomonas fluorescens*, used for the management of sedges in rice paddies.

Advantages of Bioherbicides

Bioherbicides offer several advantages over synthetic herbicides, making them an attractive option for sustainable weed management:

- **Environmental Safety:** Bioherbicides are biodegradable and pose minimal risk of soil and water contamination compared to synthetic herbicides.

- ☆ **Reduced Non-Target Effects:** Due to their host specificity, bioherbicides are less likely to harm non-target plants, beneficial organisms, or humans.
- ☆ **Resistance Management:** The use of bioherbicides can help mitigate the development of herbicide-resistant weed populations, as their modes of action differ from those of synthetic herbicides.
- ☆ **Integration with Other Practices:** Bioherbicides can be integrated into integrated weed management (IWM) systems, complementing cultural, mechanical, and chemical control methods.

Challenges in Bioherbicide Development and Use

Despite their potential, bioherbicides face several challenges that limit their widespread adoption:

1. **Narrow Host Range:** The species-specific nature of bioherbicides can be a limitation in fields with diverse weed populations.
2. **Environmental Sensitivity:** The efficacy of bioherbicides can be influenced by environmental factors such as temperature, humidity, and soil conditions, making their performance less predictable.
3. **Production and Formulation:** Scaling up the production of bioherbicides and developing stable formulations with a long shelf life can be technically and economically challenging.
4. **Regulatory Hurdles:** Bioherbicides must undergo rigorous testing and regulatory approval processes, which can be time-consuming and costly.
5. **Slow Action:** Compared to synthetic herbicides, bioherbicides often act more slowly, requiring longer periods to achieve effective weed control.

C. Recent Advances in Bioherbicide Research

Advances in biotechnology and microbial ecology are driving the development of next-generation bioherbicides. Key innovations include:

- ☆ **Genetic Engineering:** Genetic modification of microbial strains to enhance their virulence, stability, or adaptability to diverse environmental conditions.
- ☆ **Microbial Consortia:** The use of mixed microbial communities with complementary modes of action to target multiple weed species simultaneously.
- ☆ **Nanoformulations:** Incorporating bioherbicides into nanocarriers to improve their stability, delivery, and efficacy under field conditions.

☆ **Metagenomics and Bioprospecting:** High-throughput sequencing and metagenomic approaches are being used to identify novel bioherbicidal organisms and compounds from diverse ecosystems.

Application Strategies for Bioherbicides

To maximize the efficacy of bioherbicides, proper application strategies are essential. These include:

1. **Timing of Application:** Bioherbicides are most effective when applied during the early growth stages of weeds, as young plants are more susceptible to infection or phytotoxic effects.
2. **Environmental Conditions:** Optimal temperature, humidity, and soil moisture levels are critical for the survival and activity of microbial bioherbicides.
3. **Integration with Other Methods:** Combining bioherbicides with cultural practices (*e.g.*, crop rotation) or mechanical methods (*e.g.*, tillage) can enhance weed control outcomes.
4. **Carrier and Formulation:** The use of appropriate carriers (*e.g.*, water, oil, or granules) and adjuvants can improve the delivery and persistence of bioherbicides in the field.

Future Prospects

The future of bioherbicides lies in addressing current limitations and leveraging emerging technologies. Research efforts are focused on improving the efficacy, stability, and scalability of bioherbicides, as well as expanding their host range to target a broader spectrum of weeds. Additionally, the integration of bioherbicides into precision agriculture systems, supported by remote sensing and data analytics, holds promise for enhancing their adoption and impact.

D. Allelopathy Research and Applications

Allelopathy, a biological phenomenon where plants release biochemicals (allelochemicals) into the environment to influence the growth, survival, and reproduction of neighboring plants, has gained significant attention in modern weed management. This natural process offers a sustainable alternative to synthetic herbicides and provides insights into ecological interactions in agroecosystems. This section explores the principles of allelopathy, recent research developments, and its practical applications in weed management.

Principles of Allelopathy

Allelopathy involves the production and release of secondary metabolites by plants, which can affect neighboring plants either positively or negatively. These allelochemicals are released through various pathways, such as:

- **Root Exudation**: Compounds secreted by roots into the soil.
- **Volatilization**: Release of volatile organic compounds (VOCs) from leaves or stems.
- **Decomposition**: Release of allelochemicals during the decay of plant residues.
- **Leaching**: Movement of water-soluble compounds from plant tissues into the soil.

Key classes of allelochemicals include phenolics, terpenoids, alkaloids, and flavonoids. These compounds can inhibit seed germination, root elongation, or photosynthesis in target plants, thereby suppressing weed growth.

Recent Advances in Allelopathy Research

1. Identification of Novel Allelochemicals

Recent studies have identified new allelochemicals with potent herbicidal properties. For instance, sorgoleone, a hydrophobic compound produced by sorghum (Sorghum bicolor), has shown strong inhibitory effects on photosynthesis in weeds. Similarly, benzoxazinoids from maize (Zea mays) residues have demonstrated pre-emergence weed suppression capabilities (Macías *et al.*, 2019).

2. Genetic Basis of Allelopathy

Advancements in molecular biology have enabled researchers to identify genes responsible for allelochemical biosynthesis. For example, the *Bx* gene family in maize regulates the production of benzoxazinoids, while sorghum's allelopathic potential is linked to the *SbDES* gene cluster. These findings open avenues for breeding crops with enhanced allelopathic traits.

3. Microbial Interactions and Allelopathy

Soil microorganisms play a crucial role in modulating allelopathic interactions. Certain microbes can degrade allelochemicals, reducing their efficacy, while others can enhance their activity by converting them into more potent derivatives. For example, rhizosphere bacteria have been shown to amplify the herbicidal effects of phenolic acids released by wheat (Triticum aestivum).

4. Allelopathy in Cover Crops

Cover crops, such as rye (Secale cereale) and hairy vetch (Vicia villosa), have been extensively studied for their allelopathic properties. Rye residues release hydroxamic acids, which inhibit weed seed germination, while hairy vetch produces flavonoids that suppress broadleaf weeds. These findings highlight the potential of integrating allelopathic cover crops into crop rotation systems.

Applications of Allelopathy in Weed Management

1. Natural Herbicides

Allelochemicals can serve as templates for developing bioherbicides. For example:

- **Leptospermone**, derived from the Australian bottlebrush plant (Callistemon spp.), has been used to develop mesotrione, a commercial herbicide targeting broadleaf weeds.
- **Juglone**, a quinone compound from black walnut (Juglans nigra), exhibits selective herbicidal activity against annual weeds.

2. Crop Residue Management

Incorporating allelopathic crop residues into the soil can suppress weed emergence. For instance:

- Sorghum and sunflower (*Helianthus annuus*) residues release phenolic acids that inhibit weed seed germination.
- Rice (*Oryza sativa*) straw has been shown to reduce barnyard grass (*Echinochloa crus-galli*) populations when applied as mulch.

3. Intercropping Systems

Intercropping allelopathic crops with main crops can reduce weed pressure. For example:

- Intercropping maize with cowpea (*Vigna unguiculata*) has been shown to suppress *Striga hermonthica*, a parasitic weed, due to allelochemicals released by cowpea roots.
- Wheat intercropped with mustard (*Brassica juncea*) benefits from glucosinolate-derived isothiocyanates, which inhibit weed growth.

4. Allelopathic Cover Crops

The use of allelopathic cover crops is a practical strategy for weed suppression in conservation agriculture. Examples include:

- Rye cover crops, which release allelochemicals like 2-benzoxazolinone (BOA) to suppress annual weeds.
- Buckwheat (*Fagopyrum esculentum*), which produces phenolic compounds that inhibit weed seedling growth.

5. Weed Suppression in Organic Farming

Allelopathy aligns well with organic farming practices, where synthetic herbicides are prohibited. Organic farmers can leverage allelopathic crops, such as barley (*Hordeum vulgare*) and oats (*Avena sativa*), to manage weeds naturally.

Challenges and Future Directions

While allelopathy offers promising solutions for weed management, several challenges remain:

- **Variability in Field Conditions**: The efficacy of allelochemicals can be influenced by soil type, pH, and microbial activity.
- **Non-Target Effects**: Allelochemicals may affect non-target organisms, including beneficial soil microbes and neighboring crops.
- **Scalability**: Large-scale implementation of allelopathic strategies requires further research on crop compatibility and residue management.

Future research should focus on:

- Enhancing the allelopathic potential of crops through genetic engineering or selective breeding.
- Developing formulations that stabilize allelochemicals for field application.
- Investigating the synergistic effects of allelopathy and other weed control methods, such as mechanical weeding or crop rotation.

E. Integrated Pest and Weed Management (IPWM)

Integrated Pest and Weed Management (IPWM) is a holistic approach that combines multiple strategies to manage pests and weeds in agricultural systems. It emphasizes sustainable practices that reduce reliance on chemical inputs, minimize environmental impacts, and promote long-term productivity. IPWM integrates cultural, mechanical, biological, and chemical control methods, leveraging their synergistic effects to achieve effective weed and pest management. This approach is particularly relevant in modern agriculture, where resistance to herbicides and pesticides is a growing concern.

Principles of IPWM

The foundation of IPWM lies in understanding the biology and ecology of both pests and weeds. The following principles guide its implementation:

1. **Prevention**: Preventing the establishment and spread of pests and weeds is the first line of defense. This includes practices such as using certified weed-free seeds, cleaning equipment, and maintaining field hygiene.
2. **Monitoring and Thresholds**: Regular field scouting and monitoring help identify pest and weed populations. Economic thresholds are used to determine the need for intervention, ensuring that control measures are applied only when necessary.

3. **Diversity in Control Methods**: IPWM relies on a combination of control methods to reduce the risk of resistance development and enhance overall effectiveness.
4. **Sustainability**: The approach prioritizes practices that are environmentally friendly and economically viable, ensuring long-term agricultural productivity.

Components of IPWM

1. Cultural Control

Cultural control involves modifying farming practices to suppress weed and pest populations. These practices are preventive and aim to create unfavorable conditions for their growth and reproduction.

- **Crop Rotation**: Rotating crops with different growth habits disrupts the life cycles of pests and weeds. For example, alternating cereals with legumes can reduce the prevalence of specific weed species.
- **Cover Crops**: Planting cover crops, such as clover or rye, can suppress weed growth by outcompeting them for resources and reducing soil erosion.
- **Planting Density and Timing**: Adjusting planting density and sowing dates can give crops a competitive advantage over weeds and reduce pest infestations.

2. Mechanical and Physical Control

Mechanical and physical methods involve the use of tools, machinery, or physical barriers to manage weeds and pests.

- **Tillage**: Tillage disrupts weed seedbeds and exposes seeds to desiccation or predation. However, excessive tillage can lead to soil degradation, so conservation tillage practices are often preferred.
- **Mulching**: Organic or synthetic mulches suppress weed growth by blocking sunlight and reducing soil temperature fluctuations.
- **Hand Weeding and Hoeing**: These traditional methods are labor-intensive but effective for small-scale farming systems.
- **Flame Weeding**: Propane-fueled flame weeders are used to control weeds in organic farming systems without harming the soil.

3. Biological Control

Biological control leverages natural enemies of pests and weeds, such as predators, parasitoids, and pathogens, to reduce their populations.

- **Weed-Feeding Insects**: Certain insects, such as the stem-boring weevil (*Ceutorhynchus scrobicollis*), are used to control invasive weeds like Canada thistle.

- **Pathogens**: Fungal pathogens like *Colletotrichum gloeosporioides* have been developed as bioherbicides to target specific weed species.
- **Conservation of Natural Enemies**: Preserving habitats for beneficial organisms, such as lady beetles and ground beetles, enhances biological control.

4. Chemical Control

Chemical control remains an integral part of IPWM but is used judiciously to minimize environmental and health risks.

- **Selective Herbicides**: These target specific weed species without harming crops. For example, glyphosate is widely used for controlling broad-spectrum weeds in no-till systems.
- **Herbicide Rotation and Mixtures**: Rotating herbicides with different modes of action prevents the development of resistance. Combining herbicides can also enhance efficacy.
- **Precision Application**: Advances in technology, such as GPS-guided sprayers and drone-based applications, enable precise herbicide application, reducing wastage and off-target effects.

5. Genetic Control

Genetic control involves the use of crop varieties that are resistant to pests and weeds.

- **Herbicide-Tolerant Crops**: Genetically modified crops, such as glyphosate-resistant soybeans, allow for effective weed control without damaging the crop.
- **Pest-Resistant Varieties**: Breeding crops with resistance to specific pests, such as Bt cotton, reduces the need for chemical pesticides.

Emerging Technologies in IPWM

Advancements in technology are revolutionizing IPWM, making it more precise and efficient.

- **Remote Sensing and Drones**: These tools enable real-time monitoring of weed and pest populations, allowing for targeted interventions.
- **Artificial Intelligence (AI)**: AI-powered systems analyze field data to predict pest and weed outbreaks and recommend optimal control measures.
- **Robotics**: Autonomous robots equipped with sensors and mechanical weeders are being developed for precision weed management.

Challenges in IPWM Implementation

Despite its benefits, IPWM faces several challenges:

- **Resistance Management**: The overuse of chemical controls can lead to resistance in both weeds and pests, necessitating constant innovation in control strategies.
- **Knowledge and Training**: Effective implementation requires a thorough understanding of pest and weed biology, which may be lacking among some farmers.
- **Economic Constraints**: The initial costs of adopting IPWM practices, such as purchasing equipment or biocontrol agents, can be prohibitive for small-scale farmers.

Case Study: Successful IPWM Implementation

A notable example of IPWM success is the management of Palmer amaranth (*Amaranthus palmeri*) in U.S. cotton fields. This weed developed resistance to glyphosate, prompting the adoption of IPWM strategies. Farmers integrated cultural practices like cover cropping, mechanical methods such as hand-weeding, and herbicide rotation to control its spread. This multi-pronged approach not only reduced weed pressure but also improved soil health and biodiversity (Norsworthy *et al.*, 2012).

Conclusion

Integrated Pest and Weed Management is a dynamic and sustainable approach to modern agriculture. By combining diverse control methods and leveraging technological advancements, IPWM addresses the challenges of pest and weed management while promoting environmental stewardship and economic viability.

Chapter 11

Advanced Chemical Control Strategies

Advanced chemical control strategies involve the precise application of herbicides to target weed populations while minimizing environmental impact and promoting sustainable agriculture. These strategies integrate cutting-edge technologies, innovative formulations, and tailored application techniques to enhance efficacy and reduce herbicide resistance.

1. Site-Specific Herbicide Application

- ✰ Precision agriculture tools, such as GPS-guided systems and drones, enable site-specific herbicide application. These technologies allow farmers to apply herbicides only where weeds are detected, reducing chemical usage and costs.
- ✰ Weed detection systems using machine learning and hyperspectral imaging can identify weed species and densities, enabling targeted treatments (Shaner, 2014).

2. Herbicide Formulation Innovations

- ✰ Nano-formulated herbicides enhance the delivery and absorption of active ingredients, increasing their effectiveness at lower doses. These formulations reduce environmental contamination and improve weed control in challenging conditions.
- ✰ Controlled-release herbicides, encapsulated in biodegradable polymers, ensure gradual release of active ingredients, reducing leaching and volatilization.

3. Herbicide Resistance Management

- ✰ Rotating herbicides with different modes of action prevents the selection of resistant weed biotypes. For example, alternating between acetolactate synthase (ALS) inhibitors and synthetic auxins can delay resistance development.

- Tank mixing herbicides with complementary modes of action is another effective strategy to combat resistance (Heap, 2023).

4. Integrated Chemical and Biological Approaches

- Combining chemical herbicides with bioherbicides, such as plant-derived allelopathic compounds or microbial formulations, can enhance weed suppression while reducing reliance on synthetic chemicals.

A. Introduction to Advanced Chemical Control Strategies

Chemical weed control has long been a cornerstone of modern agriculture, offering a practical and efficient method for managing weed populations. However, with the increasing complexity of agricultural systems, the emergence of herbicide resistance, and growing environmental concerns, traditional chemical control methods are no longer sufficient. Advanced chemical control strategies have emerged as a solution, integrating cutting-edge technologies, precision application techniques, and novel herbicide formulations to address these challenges. This section explores the principles, innovations, and applications of advanced chemical control strategies in weed management.

Principles of Advanced Chemical Control Strategies

Advanced chemical control strategies are built on the foundation of optimizing herbicide efficacy while minimizing environmental impact and the risk of resistance development. These strategies are guided by the following principles:

- **Target-Specific Action**: Modern herbicides are designed to target specific biochemical pathways unique to weeds, reducing off-target effects on crops and non-target organisms.
- **Integrated Weed Management (IWM)**: Chemical control is integrated with cultural, mechanical, and biological methods to create a holistic weed management plan.
- **Resistance Management**: Strategies such as herbicide rotation, tank mixing, and the use of multiple modes of action are employed to delay the development of herbicide-resistant weed populations.
- **Precision Application**: Advances in application technology ensure that herbicides are applied at the right time, in the right place, and at the correct dosage to maximize effectiveness and reduce waste.

Innovations in Herbicide Formulations

The development of novel herbicide formulations has significantly enhanced the efficacy and safety of chemical weed control. Key innovations include:

1. **Nano-Formulations**: Nanotechnology has enabled the creation of herbicide formulations with improved solubility, stability, and controlled release properties. For example, nano-encapsulation allows for the slow release of active ingredients, reducing the frequency of application and minimizing environmental contamination.
2. **Safeners and Adjuvants**: Safeners are compounds added to herbicides to protect crops from potential phytotoxicity. Adjuvants, on the other hand, improve the performance of herbicides by enhancing their spreadability, adhesion, and penetration into weed tissues.
3. **Herbicide-Tolerant Crops**: Advances in genetic engineering have led to the development of crops that are tolerant to specific herbicides, such as glyphosate and glufosinate. This allows for the selective control of weeds without harming the crop.
4. **Protoporphyrinogen Oxidase (PPO) Inhibitors**: A new class of herbicides targeting the PPO enzyme has been introduced to combat resistant weed species. These herbicides disrupt chlorophyll synthesis, leading to rapid weed desiccation.

Precision Application Technologies

Precision agriculture has revolutionized chemical weed control by enabling site-specific herbicide application. This approach reduces herbicide use, minimizes environmental impact, and lowers production costs. Key technologies include:

- **Weed Detection Systems**: Advanced sensors and imaging technologies, such as hyperspectral imaging and machine vision, are used to detect weeds in real-time. These systems can differentiate between crops and weeds based on spectral signatures and morphological features.
- **Variable Rate Technology (VRT)**: VRT allows for the application of herbicides at variable rates across a field, based on weed density and distribution. This ensures that only the required amount of herbicide is applied, reducing waste and environmental contamination.
- **Drones and Autonomous Sprayers**: Unmanned aerial vehicles (UAVs) and autonomous ground sprayers equipped with GPS and weed detection systems can precisely target weed infestations. These technologies are particularly useful in areas that are difficult to access with conventional equipment.

☆ **Electrostatic Spraying**: This technique charges herbicide droplets with an electric charge, ensuring better adhesion to weed surfaces and reducing drift.

Herbicide Resistance Management

The widespread use of herbicides has led to the evolution of herbicide-resistant weed populations, posing a significant challenge to chemical weed control. Advanced strategies to manage resistance include:

1. **Herbicide Rotation**: Alternating herbicides with different modes of action prevents weeds from developing resistance to a single mode of action.
2. **Tank Mixing**: Combining herbicides with complementary modes of action in a single application reduces the likelihood of resistance development.
3. **Sequential Applications**: Applying herbicides with different modes of action in sequence ensures that resistant weeds are effectively controlled.
4. **High-Dose/Low-Dose Strategies**: Adjusting herbicide dosages based on weed biology and environmental conditions can help manage resistance. High doses are used to eliminate resistant biotypes, while low doses reduce selection pressure.

Environmental Considerations

While chemical control remains an essential component of weed management, its environmental impact cannot be overlooked. Advanced strategies aim to minimize these impacts through:

☆ **Biodegradable Formulations**: The development of herbicides that degrade rapidly in the environment reduces the risk of soil and water contamination.

☆ **Buffer Zones**: Establishing buffer zones around water bodies prevents herbicide runoff and protects aquatic ecosystems.

☆ **Drift Reduction Technologies**: Nozzles and sprayers designed to reduce drift ensure that herbicides are applied only to target areas.

Case Studies and Applications

1. **Glyphosate-Resistant Palmer Amaranth**: In the United States, glyphosate-resistant Palmer amaranth (Amaranthus palmeri) has become a major concern. Farmers have adopted advanced strategies such as tank mixing PPO inhibitors with glyphosate and using precision sprayers to target infestations.

2. **Site-Specific Weed Management in Europe**: In European countries, precision agriculture technologies, including weed detection systems and VRT, have been successfully implemented to manage weeds in cereal crops, reducing herbicide use by up to 30 per cent (Kudsk *et al.*, 2022).

Future Directions

The future of advanced chemical control strategies lies in the integration of artificial intelligence (AI), robotics, and biotechnology. AI-powered decision support systems can analyze field data to recommend optimal herbicide application plans, while robotic weeders equipped with precision sprayers can autonomously manage weeds. Additionally, the development of bioherbicides derived from natural compounds offers a sustainable alternative to synthetic herbicides.

B. New-Generation Herbicides

Weed management has always been a critical aspect of modern agriculture, and the advent of new-generation herbicides has revolutionized this field. These herbicides are designed to address the limitations of traditional chemical weed control methods, such as resistance development, environmental impact, and non-target crop damage. New-generation herbicides are characterized by their novel modes of action, improved selectivity, and reduced environmental footprint. This section explores the key features, mechanisms, and applications of these advanced herbicides.

Key Features of New-Generation Herbicides

New-generation herbicides have been developed to meet the demands of sustainable and efficient weed management. Their distinguishing features include:

- **Novel Modes of Action**: Unlike older herbicides, which often target a limited number of physiological pathways, new-generation herbicides act on previously unexploited biochemical processes in weeds. This reduces the likelihood of resistance development.
- **Improved Selectivity**: These herbicides are designed to target specific weed species while minimizing damage to crops and beneficial flora. This is achieved through advancements in molecular chemistry and formulation technologies.
- **Lower Application Rates**: Many new-generation herbicides are effective at much lower doses compared to traditional products, reducing the chemical load on the environment.
- **Reduced Environmental Persistence**: These herbicides degrade more rapidly in soil and water, minimizing their long-term environmental impact.

- **Compatibility with Integrated Weed Management (IWM)**: New-generation herbicides are often designed to complement non-chemical weed control methods, such as mechanical weeding and crop rotation, as part of an integrated approach.

Modes of Action of New-Generation Herbicides

The mode of action (MOA) of a herbicide refers to the specific biochemical process it disrupts in the weed. New-generation herbicides have introduced innovative MOAs that address the challenge of herbicide resistance. Some notable examples include:

1. Inhibition of HPPD (4-Hydroxyphenylpyruvate Dioxygenase)

HPPD inhibitors are a class of herbicides that block the enzyme 4-hydroxyphenylpyruvate dioxygenase, which is essential for carotenoid biosynthesis in plants. Carotenoids protect chlorophyll from photooxidation, and their absence leads to bleaching and death of the weed. Examples include:

- **Mesotrione**: Widely used in maize crops, mesotrione is effective against broadleaf weeds and some grasses.
- **Isoxaflutole**: Known for its pre-emergence activity, isoxaflutole is used in crops like soybeans and corn.

2. Inhibition of ALS (Acetolactate Synthase)

ALS inhibitors interfere with the synthesis of branched-chain amino acids (valine, leucine, and isoleucine) in plants. These herbicides are highly selective and effective at low doses. Examples include:

- **Flucarbazone**: Used in wheat and barley for controlling grassy weeds.
- **Imazamox**: Commonly applied in legume crops for post-emergence weed control.

3. Inhibition of PPO (Protoporphyrinogen Oxidase)

PPO inhibitors disrupt chlorophyll biosynthesis by targeting the enzyme protoporphyrinogen oxidase. This leads to the accumulation of toxic intermediates, causing cell membrane damage and plant death. Examples include:

- **Saflufenacil**: Effective for broad-spectrum weed control in crops like soybeans and cereals.
- **Flumioxazin**: Used in pre-emergence applications for controlling broadleaf weeds.

4. Synthetic Auxins

Synthetic auxins mimic natural plant hormones and disrupt normal growth processes, leading to uncontrolled cell division and eventual death. These herbicides are particularly effective against broadleaf weeds. Examples include:

- **Halauxifen-methyl**: A new-generation synthetic auxin with low environmental persistence, used in cereals.
- **Florpyrauxifen-benzyl**: Designed for use in rice paddies, this herbicide targets aquatic and terrestrial weeds.

Advancements in Formulation Technologies

The efficacy and environmental safety of new-generation herbicides are further enhanced by advancements in formulation technologies. These innovations include:

- **Microencapsulation**: Active ingredients are enclosed in microscopic capsules, allowing for controlled release and reduced volatilization. This minimizes off-target effects and improves application precision.
- **Nanoformulations**: Nanotechnology is increasingly being used to improve the solubility, stability, and bioavailability of herbicides. For example, nanoemulsions of herbicides like atrazine have shown enhanced weed control efficacy at lower doses (Kah *et al.*, 2018).
- **Drift-Reduction Technologies**: New formulations are designed to reduce spray drift, ensuring that the herbicide reaches its intended target and minimizing environmental contamination.

Herbicide Resistance Management

One of the primary drivers for the development of new-generation herbicides is the growing problem of herbicide resistance. Weeds such as *Amaranthus palmeri* (Palmer amaranth) and *Lolium rigidum* (annual ryegrass) have developed resistance to multiple herbicide classes, threatening crop productivity worldwide. New-generation herbicides play a crucial role in resistance management through:

- **Rotating Modes of Action**: Using herbicides with different MOAs in rotation reduces the selection pressure on weed populations, delaying resistance development.
- **Tank Mixing**: Combining herbicides with complementary MOAs in a single application can prevent the survival of resistant weed biotypes.
- **Site-Specific Applications**: Precision agriculture technologies, such as GPS-guided sprayers and weed-detection sensors, enable targeted herbicide applications, reducing the risk of resistance.

Examples of Commercially Available New-Generation Herbicides

Several new-generation herbicides have been introduced to the market in recent years. Some notable examples include:

- **Tiafenacil**: A PPO inhibitor with rapid action against broadleaf weeds, tiafenacil is used in pre-plant and pre-emergence applications for crops like soybeans and cereals.
- **Bixlozone**: This herbicide inhibits the enzyme DOXP synthase, a novel MOA that disrupts plastid isoprenoid biosynthesis. Bixlozone is effective against a wide range of weeds in cereals and oilseeds.
- **Cyclopyrimorate**: Targeting the enzyme homogentisate solanesyltransferase (HST), cyclopyrimorate is a novel HPPD inhibitor used in rice cultivation.

Environmental and Regulatory Considerations

New-generation herbicides are subject to rigorous regulatory scrutiny to ensure their safety for humans, non-target organisms, and the environment. Key considerations include:

- **Toxicological Profiles**: These herbicides are designed to have low toxicity to mammals, birds, and aquatic organisms.
- **Residue Management**: Advanced formulations and lower application rates reduce the risk of chemical residues in harvested crops.
- **Compliance with Sustainability Goals**: Many new-generation herbicides align with global sustainability initiatives, such as the European Union's Farm to Fork Strategy, which aims to reduce pesticide use by 50 per cent by 2030.

Future Directions

The development of new-generation herbicides is an ongoing process, driven by advances in molecular biology, synthetic chemistry, and precision agriculture. Emerging trends include:

- **RNA Interference (RNAi)-Based Herbicides**: These herbicides use double-stranded RNA molecules to silence essential genes in weeds, offering a highly specific and environmentally friendly mode of action.
- **Biological Herbicides**: Derived from natural sources such as microorganisms and plant extracts, biological herbicides are gaining attention as a sustainable alternative to synthetic chemicals.
- **Digital Weed Management**: Integration of herbicides with digital tools, such as machine learning algorithms and drone-based weed mapping, is expected to enhance the precision and efficiency of weed control.

C. Controlled-Release Herbicide Formulations

Controlled-release herbicide formulations represent a significant advancement in modern weed management, offering a precise, efficient, and environmentally sustainable approach to weed control. These formulations are designed to release active ingredients at a controlled rate over a specified period, ensuring prolonged efficacy while minimizing environmental risks. This section explores the principles, benefits, types, and applications of controlled-release herbicide formulations, supported by the latest research and examples.

Principles of Controlled-Release Herbicide Formulations

Controlled-release herbicides rely on advanced delivery systems that regulate the release of active ingredients into the soil or plant environment. The release mechanism is typically governed by physical, chemical, or biological processes, such as:

- **Diffusion:** The active ingredient diffuses through a polymer coating or matrix at a controlled rate.
- **Degradation:** The release occurs as the carrier material (*e.g.*, polymer or encapsulant) degrades due to environmental factors like moisture, temperature, or microbial activity.
- **Osmotic Pressure:** A semi-permeable membrane allows water to enter, creating pressure that releases the herbicide.

These mechanisms ensure that the herbicide is available at optimal concentrations for weed control over an extended period, reducing the need for frequent applications.

Benefits of Controlled-Release Herbicide Formulations

The adoption of controlled-release formulations offers several advantages over conventional herbicide applications:

1. Enhanced Efficacy

- Controlled-release systems maintain consistent herbicide concentrations in the target area, improving weed control performance.
- They reduce the risk of herbicide leaching or volatilization, ensuring the active ingredient remains in the root zone.

2. Environmental Safety

- By minimizing off-target movement, these formulations reduce contamination of water bodies and non-target organisms.
- They lower the risk of herbicide resistance development by providing sustained, sub-lethal doses to weeds.

3. Reduced Labor and Costs

- Fewer applications are required due to the extended activity of the herbicide, saving time and labor costs.
- Lower herbicide usage reduces input costs for farmers.

4. **Improved Crop Safety**
   - ✰ Controlled-release formulations reduce the potential for crop injury caused by high concentrations of herbicides.

Types of Controlled-Release Herbicide Formulations

Several types of controlled-release systems have been developed, each with unique characteristics suited to specific agricultural needs:

1. **Encapsulation in Polymers**
   - ✰ Herbicides are encapsulated within polymer coatings or matrices that control the release rate.
   - ✰ Example: Atrazine encapsulated in biodegradable polymers has shown prolonged weed control in maize fields (Sharma *et al.*, 2021).

2. **Microencapsulation**
   - ✰ Herbicide molecules are enclosed within micro-sized capsules made of polymers or other materials.
   - ✰ These capsules release the active ingredient gradually through diffusion or degradation.
   - ✰ Example: Microencapsulated pendimethalin for pre-emergence weed control in soybean crops.

3. **Granular Formulations**
   - ✰ Herbicides are incorporated into granules that release the active ingredient slowly when exposed to moisture.
   - ✰ Example: Controlled-release granular formulations of metolachlor for use in rice paddies.

4. **Nanotechnology-Based Systems**
   - ✰ Nanoparticles or nanocarriers are used to deliver herbicides at a controlled rate.
   - ✰ These systems improve herbicide solubility, stability, and uptake by target weeds.
   - ✰ Example: Nano-encapsulated glyphosate for enhanced weed control efficacy and reduced environmental impact (Kah *et al.*, 2018).

Applications of Controlled-Release Herbicide Formulations

Controlled-release herbicides have diverse applications in modern agriculture, ranging from row crops to specialty crops. Key applications include:

Row Crops (*e.g.*, Maize, Soybean, Wheat)

- Controlled-release formulations are particularly beneficial in crops with long growing seasons, where extended weed control is critical.
- Example: Controlled-release atrazine in maize fields reduces the need for multiple post-emergence applications.

Horticultural Crops (e.g., Vegetables, Orchards)

- These formulations are used to manage weeds in high-value crops, where precision and crop safety are paramount.
- Example: Microencapsulated herbicides for weed control in apple orchards.

Rice Paddies

- Granular controlled-release herbicides are widely used in flooded rice systems to minimize herbicide runoff and leaching.
- Example: Controlled-release butachlor for effective weed control in wetland rice.

Non-Crop Areas

- Controlled-release herbicides are employed in non-crop areas like roadsides, industrial sites, and turfgrass to provide long-lasting weed control with minimal environmental impact.

Challenges and Future Directions

Despite their advantages, controlled-release herbicide formulations face certain challenges:

- **Cost:** The production of controlled-release systems can be more expensive than conventional formulations, limiting their adoption by small-scale farmers.
- **Complexity:** The design and development of controlled-release systems require advanced materials and technologies.
- **Regulatory Hurdles:** The approval process for new formulations can be time-consuming and costly.

Future research is focused on developing cost-effective, biodegradable, and environmentally friendly materials for controlled-release systems. Advances in nanotechnology and smart delivery systems are expected to further enhance the efficacy and sustainability of these formulations.

D. Spot Application Technologies

Spot application technologies represent a significant advancement in modern weed management, offering precision, efficiency, and sustainability. These technologies focus on identifying and treating weeds only where they exist, minimizing herbicide use, reducing environmental impact, and lowering production costs. This section explores the principles, tools, and applications of spot application technologies, emphasizing their role in integrated weed management systems.

Principles of Spot Application Technologies

Spot application technologies operate on the principle of site-specific weed management (SSWM). Unlike blanket herbicide applications, which treat an entire field uniformly, spot application targets only the areas infested with weeds. This approach relies on three key components:

1. **Detection**: Identifying the presence, location, and density of weeds using sensors, cameras, or manual scouting.
2. **Decision**: Determining the appropriate herbicide, dosage, and application method based on weed species, growth stage, and environmental conditions.
3. **Action**: Applying the herbicide precisely to the targeted area using advanced delivery systems.

By integrating these components, spot application technologies ensure that herbicides are used judiciously, preserving beneficial flora and reducing the risk of herbicide resistance.

Technologies Used in Spot Application

Recent advancements in agricultural engineering and digital technologies have led to the development of several tools and systems for spot application. These technologies can be broadly categorized into the following:

1. Optical Sensor-Based Systems

Optical sensors are widely used in spot application technologies to detect weeds based on their spectral properties. These systems differentiate weeds from crops by analyzing the light reflected from plant surfaces.

- **Normalized Difference Vegetation Index (NDVI)**: NDVI sensors measure the difference between near-infrared and visible light reflectance to identify vegetation. Weeds are detected based on their unique spectral signatures.
- **Examples**: The WeedSeeker® and Weed-It® systems use optical sensors to detect weeds in real time and activate nozzles for precise herbicide application.

2. Camera-Based Systems

Camera-based systems employ high-resolution cameras and image processing algorithms to identify and classify weeds. These systems are often integrated with artificial intelligence (AI) and machine learning (ML) technologies for enhanced accuracy.

Key Features

- ✰ Real-time weed detection and mapping.
- ✰ Differentiation between weed species and crops.
- ✰ Integration with GPS for georeferenced weed maps.
- ✰ **Example**: Blue River Technology's See and Spray system uses AI-powered cameras to identify weeds and apply herbicides selectively.

3. Drone-Based Spot Application

Unmanned aerial vehicles (UAVs), or drones, are increasingly being used for spot application. Equipped with cameras, sensors, and spray systems, drones can efficiently detect and treat weeds in inaccessible or uneven terrains.

Advantages

- ✰ High-resolution aerial imagery for weed detection.
- ✰ Precision spraying in small, targeted areas.
- ✰ Reduced labor and time requirements.
- ✰ **Example**: DJI Agras drones are widely used for spot spraying in precision agriculture.

4. Autonomous Robots

Autonomous robots equipped with weed detection and spraying systems represent the cutting edge of spot application technologies. These robots navigate fields independently, using GPS and AI to identify and treat weeds.

- ✰ **Example**: The ecoRobotix system uses solar-powered robots to detect and spray weeds with minimal herbicide usage.

Applications of Spot Application Technologies

Spot application technologies have diverse applications in modern agriculture, ranging from row crops to specialty crops. Key applications include:

1. Herbicide Reduction

Spot application significantly reduces the volume of herbicides used, leading to cost savings and environmental benefits. Studies have shown that spot spraying can reduce herbicide use by up to 90 per cent compared to blanket applications (Christensen *et al.*, 2022).

2. Resistance Management

By targeting specific weed patches, spot application minimizes the selection pressure for herbicide-resistant weed populations. This approach is particularly effective in managing herbicide-resistant species like Palmer amaranth (*Amaranthus palmeri*) and waterhemp (*Amaranthus tuberculatus*).

3. Organic Farming

In organic farming systems, spot application technologies can be used to apply non-chemical weed control methods, such as flame weeding or mechanical weeding, with precision.

4. High-Value Crops

Spot application is especially beneficial in high-value crops like fruits, vegetables, and vineyards, where weed control must be precise to avoid crop damage.

Advantages of Spot Application Technologies

The adoption of spot application technologies offers several advantages:

- **Environmental Sustainability**: Reduced herbicide use lowers the risk of soil and water contamination.
- **Cost Efficiency**: Precision targeting minimizes input costs, including herbicides and labor.
- **Improved Crop Health**: By preserving beneficial plants and reducing chemical exposure, spot application promotes overall field health.
- **Scalability**: These technologies can be adapted to various field sizes and crop types.

Challenges and Limitations

Despite their potential, spot application technologies face several challenges:

- **High Initial Costs**: The cost of equipment, such as sensors, cameras, and drones, can be prohibitive for small-scale farmers.
- **Complexity**: Operating and maintaining these systems require technical expertise.
- **Detection Limitations**: Dense weed infestations or overlapping canopies may reduce detection accuracy.
- **Weather Dependency**: Environmental factors, such as light conditions and wind, can affect the performance of sensors and spraying systems.

Future Prospects

The future of spot application technologies lies in further integration with AI, robotics, and data analytics. Emerging trends include:

- **Real-Time Weed Identification**: Advances in AI and ML will enable more accurate and faster weed detection.

- **Multi-Modal Systems**: Combining optical sensors, cameras, and drones for comprehensive weed management.
- **Cloud-Based Platforms**: Centralized data storage and analysis for improved decision-making.

E. Precision Agriculture and Chemical Application

Precision agriculture has revolutionized modern weed management by enabling site-specific chemical application, reducing input costs, minimizing environmental impacts, and improving overall efficiency. This approach integrates advanced technologies such as GPS, remote sensing, variable rate technology (VRT), and data analytics to optimize the use of herbicides and other agrochemicals. By tailoring chemical applications to specific field conditions, precision agriculture ensures that weeds are managed effectively while conserving resources.

Principles of Precision Agriculture in Weed Management

Precision agriculture is based on the principle of spatial and temporal variability. Fields are not uniform; they vary in terms of soil type, weed density, crop health, and nutrient availability. Traditional uniform chemical application often leads to overuse in some areas and underuse in others, resulting in inefficiencies and potential environmental harm. Precision agriculture addresses this by:

- **Mapping Field Variability**: Using GPS and remote sensing, fields are divided into management zones based on weed density, soil properties, and crop health.
- **Site-Specific Application**: Herbicides are applied only where needed, in the required amounts, and at the right time.
- **Data-Driven Decisions**: Advanced software and algorithms analyze data collected from sensors, drones, and satellites to guide chemical application.

Technologies Enabling Precision Chemical Application

1. Geographic Information Systems (GIS) and Global Positioning Systems (GPS)

GIS and GPS are foundational technologies in precision agriculture. They enable accurate mapping of weed infestations and tracking of machinery during chemical application. For example, GPS-guided sprayers can follow pre-determined routes to ensure precise herbicide delivery, reducing overlap and missed areas.

2. Remote Sensing

Remote sensing technologies, such as drones and satellites, use multispectral and hyperspectral imaging to detect weed infestations. These tools can differentiate between crops and weeds based on their spectral signatures, allowing for targeted herbicide application. For instance, weeds with higher chlorophyll content may reflect light differently than crops, enabling their identification.

3. Variable Rate Technology (VRT)

VRT allows sprayers to adjust the rate of herbicide application in real-time based on field conditions. This is achieved through sensors that measure weed density and crop health on-the-go. For example, a sprayer equipped with VRT can increase herbicide dosage in areas with high weed pressure and reduce it in areas with low weed density.

4. Machine Vision and Artificial Intelligence (AI)

Machine vision systems, combined with AI algorithms, are increasingly being used to identify weed species and determine their growth stage. These systems can guide robotic sprayers to apply herbicides with pinpoint accuracy. For example, AI-powered weed recognition systems can distinguish between broadleaf weeds and grasses, enabling selective herbicide application.

Advantages of Precision Chemical Application

1. Reduced Herbicide Usage

Precision agriculture significantly reduces the volume of herbicides required by targeting only infested areas. This not only lowers input costs but also minimizes the risk of chemical runoff and groundwater contamination.

2. Improved Weed Control Efficiency

By applying the right herbicide at the right rate and time, precision agriculture ensures effective weed control. This reduces the likelihood of herbicide resistance developing in weed populations.

3. Environmental Sustainability

Targeted application reduces the environmental footprint of chemical usage. For example, precision sprayers equipped with drift-reduction nozzles minimize off-target herbicide deposition, protecting non-target plants and organisms.

4. Enhanced Crop Yield and Profitability

By optimizing weed control, precision agriculture promotes healthier crop growth, leading to higher yields and improved profitability for farmers.

Challenges in Precision Chemical Application

Despite its advantages, precision agriculture faces several challenges:

- ☆ **High Initial Investment**: The cost of acquiring and maintaining precision agriculture equipment, such as GPS-guided sprayers and drones, can be prohibitive for small-scale farmers.

- **Data Management**: Collecting, analyzing, and interpreting large volumes of data require specialized skills and software, which may not be readily available to all farmers.
- **Technology Integration**: Integrating various technologies, such as remote sensing, VRT, and AI, into a seamless system can be complex and time-consuming.
- **Weed Identification Limitations**: While machine vision and AI have advanced significantly, they may still struggle to identify weeds in complex field conditions, such as those with overlapping vegetation.

Case Studies and Real-World Applications

1. Weed Mapping with Drones

In a study conducted in Australia, drones equipped with multispectral cameras were used to map weed infestations in wheat fields. The data collected was processed using GIS software to create weed density maps, which were then used to guide site-specific herbicide application (Lamb *et al.*, 2021).

2. AI-Guided Robotic Sprayers

A European research initiative developed robotic sprayers equipped with AI-powered weed recognition systems. These robots were able to selectively spray herbicides on individual weeds, reducing chemical usage by up to 90 per cent compared to traditional methods (Perez-Ruiz *et al.*, 2020).

Future Trends in Precision Chemical Application

1. Integration of IoT and Big Data

The Internet of Things (IoT) and big data analytics are expected to play a significant role in the future of precision agriculture. IoT-enabled sensors can provide real-time data on weed growth, soil conditions, and weather, while big data analytics can process this information to optimize chemical application strategies.

2. Autonomous Spraying Systems

Autonomous vehicles and drones are likely to become more prevalent in precision agriculture. These systems can operate independently, applying herbicides with high accuracy and efficiency.

3. Advances in AI and Machine Learning

As AI and machine learning algorithms continue to improve, they will enable more accurate weed identification and prediction of weed growth patterns. This will further enhance the precision of chemical application.

Best Practices for Implementing Precision Chemical Application

To maximize the benefits of precision agriculture, the following best practices should be adopted:

- Conduct regular field assessments to update weed density maps and management zones.
- Use calibrated equipment to ensure accurate herbicide delivery.
- Train operators in the use of precision agriculture technologies and data interpretation.
- Monitor and evaluate the effectiveness of chemical applications to refine strategies over time.

F. Nano-Enhanced Herbicides for Increased Efficiency

The advent of nanotechnology in agriculture has revolutionized various aspects of crop production, including weed management. Nano-enhanced herbicides represent a cutting-edge innovation, offering improved efficacy, reduced environmental impact, and enhanced precision in weed control. This section explores the principles, mechanisms, and applications of nano-enhanced herbicides, providing agricultural professionals and researchers with a comprehensive understanding of this emerging technology.

Principles of Nano-Enhanced Herbicides

Nano-enhanced herbicides utilize nanotechnology to improve the delivery, efficacy, and environmental safety of herbicidal compounds. These formulations typically involve the encapsulation or conjugation of active herbicidal ingredients with nanoparticles, which serve as carriers. The nanoscale size of these carriers (1–100 nm) enhances the bioavailability and targeted delivery of herbicides, reducing wastage and off-target effects.

Key principles include:

- **Controlled Release**: Nanoparticles enable the slow and controlled release of herbicides, ensuring prolonged activity and reducing the need for frequent applications.
- **Targeted Delivery**: Functionalized nanoparticles can be engineered to specifically bind to weed tissues, minimizing damage to non-target plants.
- **Enhanced Solubility**: Nanocarriers improve the solubility of poorly water-soluble herbicides, increasing their bioavailability and effectiveness.
- **Reduced Dosage**: Due to their high efficiency, nano-enhanced herbicides require lower dosages compared to conventional formulations, reducing chemical load on the environment.

Types of Nanoparticles Used in Herbicides

Several types of nanoparticles are employed in the formulation of nano-enhanced herbicides, each offering unique properties:

1. Polymeric Nanoparticles

- Made from biodegradable polymers such as polylactic acid (PLA) or polycaprolactone (PCL).
- Provide controlled release of herbicides over time.
- Example: Encapsulation of atrazine in polymeric nanoparticles has demonstrated prolonged weed suppression (Kah *et al.*, 2018).

2. Lipid-Based Nanoparticles

- Include liposomes and solid lipid nanoparticles.
- Enhance the solubility and stability of hydrophobic herbicides.
- Example: Liposome-encapsulated glyphosate formulations have shown improved uptake by weed tissues.

3. Metallic Nanoparticles

- Composed of metals like silver, zinc, or iron.
- Exhibit antimicrobial properties that can complement weed control.
- Example: Zinc oxide nanoparticles have been used to enhance the efficacy of 2,4-D herbicide.

4. Silica-Based Nanoparticles

- Offer high surface area and tunable porosity.
- Facilitate the adsorption and controlled release of herbicides.
- Example: Mesoporous silica nanoparticles loaded with paraquat have demonstrated sustained herbicidal activity.

Mechanisms of Action

Nano-enhanced herbicides function through several mechanisms that improve their overall performance:

- **Improved Penetration**: The small size of nanoparticles allows them to penetrate plant cuticles and cell walls more effectively, ensuring better absorption of the active ingredient.
- **Increased Retention**: Nanoparticles adhere to leaf surfaces more effectively, reducing runoff and enhancing herbicide retention.
- **Site-Specific Action**: Functionalized nanoparticles can be designed to release herbicides in response to specific environmental triggers, such as pH or temperature, ensuring site-specific action.

- **Synergistic Effects**: Some nanoparticles, such as metallic ones, exhibit inherent bioactivity that can enhance the herbicidal effect.

Advantages of Nano-Enhanced Herbicides

The adoption of nano-enhanced herbicides offers several advantages over conventional formulations:

Environmental Safety

- Reduced chemical runoff and leaching into soil and water bodies.
- Lower risk of non-target toxicity to beneficial organisms.

Economic Benefits

- Reduced herbicide usage leads to cost savings for farmers.
- Prolonged activity reduces the frequency of application.

Improved Weed Control

- Enhanced efficacy against herbicide-resistant weed species.
- Better performance under adverse environmental conditions, such as drought.

Sustainability

- Aligns with integrated weed management (IWM) principles by minimizing chemical inputs.

Challenges and Limitations

Despite their potential, nano-enhanced herbicides face several challenges that must be addressed for widespread adoption:

Regulatory Hurdles

- The lack of standardized guidelines for the approval and use of nanomaterials in agriculture.
- Concerns over the long-term environmental and human health impacts of nanoparticles.

Cost of Production

- High production costs of nanoparticles may limit their affordability for small-scale farmers.

Knowledge Gap

- Limited awareness and technical expertise among farmers regarding the use of nano-enhanced herbicides.

Environmental Persistence

- Potential accumulation of non-biodegradable nanoparticles in the environment.

Recent Advances and Applications

Recent research has demonstrated the potential of nano-enhanced herbicides to address pressing challenges in weed management:

Smart Herbicide Delivery Systems

- Development of pH-responsive nanoparticles that release herbicides in alkaline environments, such as the rhizosphere of certain weeds.
- Example: Chitosan-based nanoparticles loaded with glufosinate have shown selective action against broadleaf weeds.

Combination Formulations

- Integration of nano-enhanced herbicides with other weed control methods, such as bioherbicides or mechanical weeding, for synergistic effects.

Field Trials

- Large-scale field trials of nano-formulated herbicides have demonstrated significant reductions in weed biomass and improved crop yields.

Future Prospects

The future of nano-enhanced herbicides lies in the development of more sustainable and cost-effective formulations. Advances in nanotechnology, such as green synthesis of nanoparticles using plant extracts, hold promise for reducing production costs and environmental impact. Additionally, interdisciplinary research involving agronomy, materials science, and environmental science will be crucial for overcoming existing challenges and unlocking the full potential of this technology.

G. Reduced Environmental Impact of Advanced Chemical Strategies

The evolution of chemical weed control methods has significantly enhanced agricultural productivity. However, traditional herbicide use has often raised concerns regarding environmental sustainability, including soil degradation, water contamination, and harm to non-target organisms. Advanced chemical strategies aim to mitigate these issues by leveraging precision application technologies, novel formulations, and environmentally benign herbicides. This section explores how these innovations reduce the environmental footprint of chemical weed management.

1. Precision Application Technologies

Precision application technologies have revolutionized herbicide use by ensuring targeted delivery, minimizing waste, and reducing off-target impacts. These technologies integrate advanced tools such as GPS-guided sprayers, variable-rate applicators, and artificial intelligence (AI)-driven weed detection systems.

- ☆ **GPS-Guided Sprayers**: These systems use satellite navigation to apply herbicides with pinpoint accuracy. By mapping field variability, they ensure that herbicides are applied only where needed, reducing over-application and runoff risks.
- ☆ **AI-Driven Weed Detection**: Machine learning algorithms and imaging technologies enable real-time weed identification. For instance, camera-equipped sprayers can distinguish between crops and weeds, applying herbicides exclusively to the latter. This approach significantly reduces herbicide usage and environmental exposure.
- ☆ **Variable-Rate Applicators**: These systems adjust the herbicide dosage based on weed density and type. By tailoring application rates, they prevent overuse and minimize chemical residues in the soil and water.

A study by Pérez-Ruiz *et al.* (2020) demonstrated that precision spraying technologies could reduce herbicide use by up to 70 per cent without compromising weed control efficacy. This reduction directly translates to lower environmental contamination and improved sustainability.

2. Development of Environmentally Benign Herbicides

The development of herbicides with improved environmental profiles is another critical advancement. These herbicides are designed to degrade rapidly, exhibit low toxicity to non-target organisms, and reduce the risk of bioaccumulation.

- ☆ **Biodegradable Herbicides**: Modern formulations prioritize rapid degradation in soil and water. For example, herbicides based on natural compounds, such as pelargonic acid and allelopathic plant extracts, break down into harmless byproducts, minimizing long-term environmental impact.
- ☆ **Low-Volatility Formulations**: Volatile herbicides can drift during application, affecting nearby ecosystems. Low-volatility formulations, such as encapsulated or water-based herbicides, reduce the risk of drift and off-target damage.
- ☆ **Selective Herbicides**: Advances in molecular biology have enabled the development of herbicides that target specific biochemical pathways

unique to weeds. For instance, herbicides targeting acetolactate synthase (ALS) inhibitors are highly selective, leaving crops and beneficial organisms unharmed.

Research by Duke (2020) highlights that the integration of natural product-based herbicides could serve as a sustainable alternative to synthetic chemicals, reducing ecological risks while maintaining weed control efficiency.

3. Controlled-Release and Nanoformulations

Controlled-release and nanoformulation technologies have emerged as game-changers in herbicide delivery. These innovations ensure that herbicides are released gradually and precisely, optimizing their efficacy while minimizing environmental exposure.

- **Controlled-Release Herbicides**: Encapsulation techniques, such as polymer-coated granules, allow herbicides to be released over an extended period. This reduces the frequency of application and prevents leaching into groundwater.
- **Nanotechnology in Herbicides**: Nanoformulations enhance the solubility, stability, and bioavailability of herbicides. For example, herbicides encapsulated in nanoparticles can penetrate weed tissues more effectively, enabling lower application rates. Additionally, nanoparticles can be engineered to degrade under specific environmental conditions, further reducing residual impact.

Studies have shown that nanoformulated herbicides can achieve up to a 50 per cent reduction in application rates while maintaining comparable weed control efficacy (Kah *et al.*, 2018). This not only reduces chemical input but also minimizes the risk of environmental contamination.

4. Integrated Weed Management (IWM) with Chemical Strategies

Integrating advanced chemical strategies into broader weed management frameworks, such as Integrated Weed Management (IWM), enhances their environmental benefits. IWM combines chemical, cultural, mechanical, and biological control methods to achieve sustainable weed control.

- **Herbicide Rotation**: Alternating herbicides with different modes of action prevents resistance development and reduces the need for higher chemical doses.
- **Spot Treatments**: Instead of blanket applications, herbicides are applied only to weed-infested areas, reducing overall chemical use.
- **Synergistic Approaches**: Combining chemical herbicides with biological agents, such as mycoherbicides, can enhance efficacy while lowering chemical dependency.

For example, the integration of precision spraying with cultural practices like cover cropping has been shown to reduce herbicide use by up to 40 per cent in certain cropping systems (Liebman *et al.*, 2016).

5. Regulatory and Industry Initiatives

Regulatory frameworks and industry initiatives play a pivotal role in promoting environmentally friendly herbicide practices. Governments and organizations are encouraging the adoption of advanced chemical strategies through incentives, certifications, and research funding.

- ☆ **Regulatory Approvals**: Many countries now require herbicides to meet stringent environmental safety standards before approval. This has driven innovation in low-toxicity and biodegradable formulations.
- ☆ **Stewardship Programs**: Industry-led programs educate farmers on the proper use of advanced technologies, ensuring compliance with best practices and minimizing environmental risks.

Chapter 12

Non-Chemical and Sustainable Approaches

Non-chemical weed management methods are integral to sustainable agriculture, reducing dependency on herbicides while minimizing environmental and health risks. These approaches emphasize ecological balance, long-term soil health, and biodiversity conservation.

1. Cultural Practices

- **Crop Rotation**: Alternating crops with different growth habits disrupts weed life cycles. For instance, rotating cereals with legumes can suppress specific weed species.
- **Cover Crops**: Fast-growing cover crops like clover or rye outcompete weeds for light, nutrients, and space. They also improve soil organic matter and reduce erosion.
- **Optimal Planting Density**: Adjusting crop density ensures rapid canopy closure, limiting light availability for weeds.

2. Mechanical Control

- **Tillage**: Implements such as rotary hoes or harrows physically uproot weeds. However, excessive tillage can degrade soil structure.
- **Mowing and Mulching**: Mowing prevents weed seed production, while organic mulches suppress weed emergence by blocking sunlight.

3. Biological Control

- **Natural Enemies**: Introducing weed-specific pathogens or insects, such as the use of *Puccinia chondrillina* fungus against skeleton weed, can effectively suppress invasive species.
- **Grazing**: Livestock, like sheep or goats, can selectively graze on weeds, reducing their biomass.

4. Thermal Methods

- **Flame Weeding**: Propane-fueled burners expose weeds to high temperatures, causing cellular damage. This method is particularly effective for small-scale organic farming systems.

5. Integrated Weed Management (IWM)

Combining multiple non-chemical methods enhances weed control efficacy while reducing environmental impacts. For example, integrating cover crops with grazing and mechanical weeding creates a synergistic effect.

6. Advances in Technology

- **Robotics and AI**: Autonomous weeders equipped with machine vision can identify and remove weeds with precision, reducing labor costs and soil disturbance.
- **Solarization**: Covering soil with transparent plastic traps solar heat, killing weed seeds and pathogens in the upper soil layers.

Non-chemical strategies align with sustainable agricultural goals, promoting resilience against herbicide resistance and fostering ecological harmony (Liebman *et al.*, 2001). These methods are increasingly relevant in organic farming systems and regions with stringent pesticide regulations.

A. Electro-Weeding Technology

Electro-weeding technology is an innovative, non-chemical weed management method that uses electrical energy to damage or destroy unwanted plants. This technology has gained attention in recent years due to its potential to reduce reliance on herbicides, address herbicide resistance, and offer an environmentally sustainable alternative for weed control. Electro-weeding is particularly useful in organic farming systems and areas where chemical herbicides are restricted or less effective.

Principle of Electro-Weeding

Electro-weeding operates on the principle of delivering high-voltage electrical energy to the weed's tissues. When an electrical current passes through the plant, it generates intense heat, causing thermal damage to the cells. This disrupts the plant's physiological processes, leading to desiccation and eventual death. The process is highly effective for both above-ground and below-ground plant parts, including roots, making it a comprehensive weed control solution.

Key principles include:

- **Electrical Circuit Formation**: The weed acts as a conductor, completing the electrical circuit between the applicator and the ground.
- **Thermal Damage**: The electrical current generates heat within the plant tissues, causing cell walls to rupture.
- **Systemic Impact**: The heat energy can travel through the plant's vascular system, damaging roots and preventing regrowth.

Components of Electro-Weeding Systems

Electro-weeding equipment typically consists of the following components:

1. **Power Source**: A generator or battery provides the high-voltage electricity required for weed destruction.
2. **Applicator**: A wand or electrode delivers the electrical current directly to the weed. Applicators are often designed to target specific weeds while minimizing damage to crops.
3. **Control Unit**: This regulates the voltage and current to ensure safe and effective operation.
4. **Insulation and Safety Features**: Insulated handles, automatic shut-offs, and other safety mechanisms protect the operator from electrical hazards.

Advantages of Electro-Weeding

Electro-weeding offers several advantages over traditional weed control methods:

- **Non-Chemical Approach**: It eliminates the need for herbicides, making it suitable for organic farming and environmentally sensitive areas.
- **Effective on Herbicide-Resistant Weeds**: Electro-weeding can control weeds that have developed resistance to chemical herbicides.
- **Broad-Spectrum Control**: It is effective against a wide range of weed species, including perennial weeds with extensive root systems.
- **Reduced Soil Disturbance**: Unlike mechanical weeding, electro-weeding does not disturb the soil, reducing the risk of erosion and preserving soil structure.
- **Precision Targeting**: Modern electro-weeding systems allow for precise application, minimizing damage to crops and non-target plants.

Limitations and Challenges

Despite its advantages, electro-weeding has certain limitations and challenges:

- ☆ **High Energy Requirements**: The process requires significant electrical energy, which may limit its scalability in large agricultural operations.
- ☆ **Cost of Equipment**: Initial investment in electro-weeding equipment can be high, making it less accessible for small-scale farmers.
- ☆ **Safety Concerns**: The use of high-voltage electricity poses safety risks to operators, requiring proper training and protective measures.
- ☆ **Weather Dependency**: Wet conditions can reduce the effectiveness of electro-weeding, as moisture can dissipate the electrical energy.

Applications in Agriculture

Electro-weeding technology has been successfully applied in various agricultural settings:

- ☆ **Row Crops**: It is commonly used in row crops such as maize, soybeans, and sugar beets to control weeds between rows without harming the crop.
- ☆ **Perennial Weed Management**: Electro-weeding is particularly effective for managing perennial weeds like Canada thistle and bindweed, as it targets both the above-ground and below-ground parts of the plant.
- ☆ **Organic Farming**: Organic farmers use electro-weeding to maintain weed-free fields without compromising organic certification standards.
- ☆ **Specialty Crops**: In high-value crops like vineyards and orchards, electro-weeding provides a precise and non-invasive weed control option.

Recent Advances in Electro-Weeding Technology

Recent advancements have improved the efficiency and usability of electro-weeding systems:

- ☆ **Automation and Robotics**: Automated and robotic electro-weeding machines are being developed to reduce labor requirements and improve precision.
- ☆ **Energy Efficiency**: Innovations in power supply technology, such as the use of renewable energy sources, are reducing the energy footprint of electro-weeding systems.
- ☆ **Integration with Precision Agriculture**: Electro-weeding is being integrated with precision agriculture tools, such as GPS and sensors, to target weeds more effectively and reduce energy consumption.

Case Studies and Research

Research studies have demonstrated the effectiveness of electro-weeding in various agricultural contexts. For example, a study conducted by Vigneau *et al.* (2021) found that electro-weeding achieved a 90 per cent reduction in weed biomass in sugar beet fields, with minimal impact on crop yield. Similarly, trials in vineyards have shown that electro-weeding can effectively control weeds while preserving soil health and biodiversity.

Safety Considerations

To ensure safe operation, the following precautions should be taken:

- Operators must wear insulated gloves and protective clothing.
- Equipment should be regularly inspected for damage or wear.
- Training programs should be implemented to educate users on proper handling and emergency procedures.

Future Prospects

As the demand for sustainable weed management solutions grows, electro-weeding technology is expected to play an increasingly important role in modern agriculture. Ongoing research and development efforts aim to address its current limitations and expand its applicability to a wider range of crops and farming systems.

B. Thermal Weed Control

Thermal weed control is an innovative and environmentally friendly method of managing weeds by applying heat to disrupt or destroy plant tissues. This technique is gaining attention as a sustainable alternative to chemical herbicides, particularly in organic farming systems and areas where herbicide resistance is a concern. By leveraging heat as a physical control method, thermal weed control minimizes the reliance on synthetic chemicals, reduces environmental contamination, and aligns with integrated weed management (IWM) principles.

Principle of Thermal Weed Control

Thermal weed control operates on the principle of exposing weeds to high temperatures, which causes rapid denaturation of proteins, cell membrane disruption, and eventual plant desiccation. The heat damages the plant's cellular structure, leading to dehydration and death. Unlike chemical herbicides, thermal methods do not rely on systemic action; instead, they target the above-ground parts of the plant.

The effectiveness of thermal weed control depends on several factors, including:

- **Weed Growth Stage**: Younger weeds with tender tissues are more susceptible to heat damage than mature plants with established root systems.
- **Heat Intensity and Duration**: Higher temperatures and prolonged exposure increase the likelihood of successful weed control.
- **Weed Species**: Different species exhibit varying levels of tolerance to heat, with broadleaf weeds generally being more vulnerable than grasses.

Methods of Thermal Weed Control

Several thermal weed control methods are available, each with specific applications and limitations. These include:

1. Flaming

Flaming involves the use of propane or natural gas burners to apply direct heat to weeds. The flame raises the temperature of the plant tissue to approximately 70–100°C, causing cell rupture and plant death. Flaming is particularly effective for controlling annual weeds in crop rows, pathways, and non-crop areas.

Advantages

- Quick and effective for small-scale operations.
- No residual effects on soil or crops.
- Suitable for organic farming systems.

Limitations

- Ineffective against perennial weeds with extensive root systems.
- Risk of fire hazards in dry conditions.

2. Hot Water and Steam

Hot water or steam weed control involves applying water heated to 95–100°C directly to weeds. The heat penetrates the plant tissue, causing thermal shock and subsequent desiccation. This method is commonly used in urban areas, such as sidewalks, parks, and gardens, where chemical herbicides are restricted.

Advantages

- Safe for use in public spaces.
- Effective for controlling weeds in hard-to-reach areas.

Limitations

- High energy and water requirements.
- Limited scalability for large agricultural fields.

3. Infrared Radiation

Infrared weed control uses radiant heat generated by infrared emitters to kill weeds. The heat is directed onto the plant surface, causing thermal injury without producing an open flame. This method is often used in precision agriculture and for weed control in greenhouses.

Advantages

- Precise application with minimal risk of collateral damage.
- Suitable for sensitive environments.

Limitations

- Slower treatment time compared to flaming.
- High initial equipment costs.

4. Microwaves

Microwave weed control is an emerging technology that uses electromagnetic waves to generate heat within the plant tissue. The microwaves cause water molecules in the plant cells to vibrate, producing heat that destroys the cells. This method is still under development but shows promise for non-chemical weed management.

Advantages

- Potential for deep soil penetration to target weed seeds.
- No chemical residues.

Limitations

- High energy consumption.
- Limited commercial availability.

Factors Influencing Thermal Weed Control Efficiency

The success of thermal weed control depends on a combination of environmental, operational, and biological factors:

- **Environmental Conditions**: Wind, humidity, and soil moisture can affect heat transfer and treatment efficacy. For example, high humidity may reduce the effectiveness of flaming by dissipating heat.
- **Weed Density**: Dense weed populations may require multiple passes or higher heat intensity to achieve satisfactory control.

- **Crop Sensitivity**: In crop systems, care must be taken to avoid damaging desirable plants. Shielded burners or precision tools can help minimize crop injury.

Applications in Modern Agriculture

Thermal weed control is increasingly being integrated into modern agricultural systems for various purposes:

- **Organic Farming**: Provides a non-chemical weed control option that aligns with organic certification standards.
- **Pre-Emergence Weed Control**: Flaming or infrared radiation can be used to eliminate weed seedlings before crop emergence, reducing competition during critical growth stages.
- **Post-Harvest Weed Management**: Thermal methods can be employed to manage weeds in fallow fields or after harvest to prepare for the next cropping season.

Challenges and Future Prospects

While thermal weed control offers numerous benefits, it also faces challenges such as high energy costs, limited effectiveness against perennial weeds, and labor-intensive operations. However, advancements in technology, such as the development of automated thermal weeders and renewable energy-powered systems, are addressing these limitations.

Research is ongoing to improve the efficiency and scalability of thermal weed control methods. For instance, studies have explored the use of solar-powered steam weeders and hybrid systems combining thermal and mechanical control (Ascard, 1995). These innovations hold promise for making thermal weed control a more viable option for large-scale agriculture.

C. Weed Seedbank Management

Weed seedbanks, defined as the reservoir of viable weed seeds present in the soil, are a critical component of weed population dynamics in agricultural systems. Effective weed seedbank management is essential for long-term weed control and sustainable crop production. This section explores the principles, strategies, and modern approaches to managing weed seedbanks, providing agricultural professionals, researchers, and students with actionable insights.

Understanding the Weed Seedbank

The weed seedbank is the primary source of weed infestations in agricultural fields. It consists of seeds deposited over time through natural dispersal, crop harvesting, and soil disturbance. Weed seeds can remain dormant in the soil for years, creating a persistent challenge for farmers.

- **Seed Longevity:** Weed seeds exhibit varying degrees of dormancy and longevity. For example, seeds of species like *Amaranthus retroflexus* (redroot pigweed) can remain viable for up to 40 years, while others, such as *Setaria viridis* (green foxtail), persist for shorter periods.
- **Seedbank Dynamics:** The size and composition of the seedbank are influenced by factors such as weed seed production, predation, decay, and germination. These dynamics are further affected by agricultural practices, soil conditions, and environmental factors.

Principles of Weed Seedbank Management

Weed seedbank management is based on the principle of reducing seed inputs and accelerating seed losses. This involves minimizing the addition of new seeds to the soil while promoting the depletion of existing seeds through germination, predation, and decay.

Key Principles

1. *Prevention of Seed Production*

- Preventing weeds from reaching reproductive maturity is critical to curbing seedbank replenishment.
- Techniques such as timely mechanical weeding, herbicide application, and crop competition can suppress seed production.

2. *Seedbank Depletion*

- Encouraging weed seed germination followed by effective control measures can deplete the seedbank.
- Practices like shallow tillage and stale seedbed techniques stimulate germination, allowing for subsequent control.

3. *Integrated Weed Management (IWM)*

- Combining cultural, mechanical, biological, and chemical control methods enhances the effectiveness of seedbank management.
- IWM approaches reduce reliance on any single method, mitigating the risk of herbicide resistance and environmental degradation.

Modern Strategies for Weed Seedbank Management

Advancements in agricultural science have introduced innovative strategies for managing weed seedbanks. These methods leverage ecological principles, precision agriculture, and emerging technologies.

1. Cultural Practices

Cultural practices aim to create unfavorable conditions for weed seed survival and germination.

Crop Rotation

- Diverse crop rotations disrupt weed life cycles by altering planting schedules, canopy structures, and soil conditions.
- For example, rotating cereals with legumes can suppress grass weed species like *Avena fatua* (wild oats).

Cover Cropping

- Cover crops, such as rye (*Secale cereale*) and clover (*Trifolium spp.*), suppress weed growth by shading the soil and releasing allelopathic compounds.
- These crops also enhance soil health, indirectly reducing seedbank persistence.

Mulching

- Organic and synthetic mulches create a physical barrier that prevents weed seed germination and establishment.
- Straw mulch, for instance, has been shown to reduce the emergence of small-seeded weeds like *Chenopodium album* (common lambsquarters).

2. Mechanical Practices

Mechanical interventions target weed seeds directly or modify soil conditions to influence seedbank dynamics.

Tillage Management

- Shallow tillage can stimulate weed seed germination, while deep tillage buries seeds beyond their germination depth.
- However, excessive tillage may lead to soil erosion and loss of organic matter, necessitating a balanced approach.

Harvest Weed Seed Control (HWSC)

- HWSC techniques, such as chaff carts and seed destructors, capture or destroy weed seeds during harvest.
- Research has shown that HWSC can reduce seedbank inputs by up to 90 per cent for species like *Lolium rigidum* (annual ryegrass) [Walsh *et al.*, 2013].

3. Biological Control

Biological control methods utilize natural enemies of weed seeds, such as seed predators and pathogens.

Seed Predation

- Ground beetles (*Carabidae*) and rodents are effective seed predators that can significantly reduce seedbank densities.
- Enhancing field margins and maintaining habitat diversity can promote these beneficial organisms.

Pathogen-Based Control

- Soilborne pathogens, such as fungi, can infect and degrade weed seeds, reducing their viability.
- For example, *Fusarium spp.* have been investigated for their potential to suppress seeds of *Abutilon theophrasti* (velvetleaf).

4. Chemical Control

Herbicides play a role in seedbank management by preventing seed production and depleting the seedbank.

Pre-Emergence Herbicides

- These herbicides inhibit weed seed germination and emergence, reducing seedbank inputs.
- For instance, atrazine is effective against broadleaf weeds in maize fields.

Residual Herbicides

- Residual herbicides provide extended control by persisting in the soil and targeting germinating seeds.
- However, their use must be carefully managed to avoid environmental contamination and resistance development.

5. Precision Agriculture

Precision agriculture technologies enable targeted and efficient weed seedbank management.

Soil Mapping

- Geospatial tools, such as GPS and GIS, allow for the identification of weed seedbank hotspots.
- Site-specific management practices can then be implemented to address these areas.

Robotics and Automation

- Autonomous weeders equipped with sensors and cameras can detect and remove weeds at early growth stages.

- These systems reduce the reliance on herbicides and labor-intensive practices.

Monitoring and Evaluation

Regular monitoring of the weed seedbank is essential for assessing the effectiveness of management strategies and making informed decisions.

Soil Sampling

- Soil samples can be collected and analyzed to estimate seedbank density and composition.
- Techniques like seed extraction and germination assays are commonly used for this purpose.

Long-Term Studies

- Long-term field experiments provide valuable insights into the impact of management practices on seedbank dynamics.
- Data from such studies can guide the development of sustainable weed control programs.

Challenges and Future Directions

Despite significant advancements, several challenges remain in weed seedbank management:

Herbicide Resistance

- The evolution of herbicide-resistant weed populations complicates chemical control efforts.
- Integrated approaches are necessary to mitigate resistance risks.

Climate Change

- Changing climatic conditions may alter weed seedbank dynamics, necessitating adaptive management strategies.

Knowledge Gaps

- Further research is needed to understand the interactions between weed seedbanks, soil health, and cropping systems.

Emerging technologies, such as machine learning and gene editing, hold promise for addressing these challenges. For instance, CRISPR-based techniques could potentially target weed seed dormancy mechanisms, offering novel solutions for seedbank management.

D. Permaculture and Agroecological Approaches

Permaculture and agroecological approaches represent holistic strategies for sustainable agriculture, emphasizing ecological balance, biodiversity, and resource efficiency. These methods are increasingly being adopted as effective tools for weed management in modern farming systems. By leveraging natural processes and designing resilient ecosystems, these approaches minimize reliance on chemical herbicides while promoting long-term soil health and productivity.

Principles of Permaculture in Weed Management

Permaculture is a design philosophy that integrates natural ecosystems with agricultural systems to create self-sustaining environments. Its application in weed management focuses on prevention rather than eradication, using ecological principles to suppress weed growth.

Key principles of permaculture relevant to weed control include:

- ☆ **Diversity and Polyculture**: Planting a variety of crops reduces the dominance of specific weed species. Diverse root systems and canopy structures compete with weeds for light, nutrients, and water.
- ☆ **Mulching**: Organic mulches, such as straw, wood chips, or cover crop residues, act as physical barriers to weed emergence. Mulching also improves soil moisture retention and microbial activity.
- ☆ **Soil Health Management**: Healthy soils with balanced nutrient profiles and active microbial communities suppress weed growth by favoring crop competitiveness.
- ☆ **Edge Design and Zoning**: Strategic planting of perennial species or hedgerows along field edges reduces weed encroachment from surrounding areas.

For example, in a permaculture system, intercropping legumes with cereals can suppress weeds while fixing nitrogen in the soil, enhancing overall productivity and reducing the need for synthetic inputs.

Agroecological Approaches to Weed Management

Agroecology applies ecological principles to agricultural systems, emphasizing the interactions between plants, soil, and the surrounding environment. This approach promotes the use of natural processes to manage weeds, reduce chemical inputs, and enhance ecosystem services.

Key Agroecological Strategies for Weed Control

1. *Crop Rotation*

- ☆ Alternating crops with different growth habits disrupts weed life cycles and reduces the buildup of weed populations.

- For instance, rotating cereals with legumes or root crops prevents the dominance of specific weed species adapted to a single crop type.

2. *Cover Cropping*

- Cover crops, such as clover, rye, or vetch, suppress weeds by outcompeting them for resources and creating a dense canopy that limits light penetration.
- Certain cover crops, like rye, release allelopathic compounds that inhibit weed seed germination.

3. *Intercropping*

- Growing multiple crops simultaneously in the same field creates a more competitive environment for weeds.
- For example, maize intercropped with beans has been shown to reduce weed biomass while improving land-use efficiency.

4. *Biological Weed Control*

- Utilizing natural enemies of weeds, such as insects, fungi, or grazing animals, can help regulate weed populations.
- For example, the fungal pathogen *Puccinia chondrillina* has been used successfully to control skeleton weed (*Chondrilla juncea*) in Australia (Watson and Wymore, 2020).

5. *Reduced Tillage*

- Minimizing soil disturbance reduces the germination of weed seeds buried in the soil. Conservation tillage practices, such as no-till or strip-till, also improve soil structure and organic matter content.

6. *Weed Seed Bank Management*

- Techniques such as stale seedbed preparation, where weeds are allowed to germinate and then destroyed before planting the main crop, reduce the weed seed bank in the soil.

Integration of Permaculture and Agroecology

Permaculture and agroecology share common goals of sustainability, resilience, and reduced reliance on external inputs. Integrating these approaches can create synergistic effects in weed management. For instance:

- Combining **permaculture mulching** with **agroecological cover cropping** can enhance weed suppression while improving soil health.
- Designing **permaculture zones** with **agroecological crop rotations** ensures long-term weed control by disrupting weed life cycles and maintaining biodiversity.

Advantages of Permaculture and Agroecological Weed Management

Environmental Benefits

- Reduced chemical herbicide use minimizes soil and water contamination.
- Enhanced biodiversity supports pollinators and natural pest control agents.

Economic Benefits

- Lower input costs for herbicides and fertilizers.
- Improved soil fertility reduces the need for synthetic amendments.

Social Benefits

- Promotes farmer autonomy by reducing dependence on external inputs.
- Encourages knowledge-sharing and community-based farming practices.

Challenges and Limitations

While these approaches offer numerous benefits, they also present challenges:

- **Labor Intensity**: Practices like mulching, intercropping, and manual weed removal can be labor-intensive.
- **Knowledge Requirements**: Successful implementation requires a deep understanding of ecological interactions and site-specific conditions.
- **Transition Period**: Shifting from conventional to agroecological systems may involve a temporary decline in productivity as the ecosystem stabilizes.

Recent Advances and Research

Recent studies have highlighted the potential of integrating digital tools with agroecological practices. For example, precision agriculture technologies, such as drones and sensors, can monitor weed populations and optimize the application of agroecological strategies (Smith *et al.*, 2022). Additionally, research into allelopathic cover crops and microbial weed suppressants continues to expand the toolbox for sustainable weed management.

Conclusion

Permaculture and agroecological approaches offer innovative, sustainable solutions for weed management by leveraging natural processes and ecological principles. These methods align with the goals of modern agriculture to enhance productivity while preserving environmental integrity.

E. Role of Mulches for Weed Management

Weeds are one of the most persistent challenges in agriculture, competing with crops for essential resources such as water, nutrients, light, and space. Mulching, a widely adopted cultural practice, has emerged as an effective and sustainable weed management strategy. Mulches not only suppress weed growth but also improve soil health, conserve moisture, and regulate soil temperature. This section explores the role of mulches in weed management, their types, mechanisms of action, and practical applications in modern agriculture.

Mechanisms of Weed Suppression by Mulches

Mulches suppress weeds through several mechanisms, including:

- ☆ **Physical Barrier:** Mulches act as a physical barrier, preventing sunlight from reaching the soil surface. This inhibits the germination and growth of light-sensitive weed seeds.
- ☆ **Allelopathic Effects:** Organic mulches, such as straw or wood chips, may release allelopathic compounds during decomposition. These compounds can inhibit weed seed germination and growth.
- ☆ **Soil Temperature Regulation:** Mulches moderate soil temperature, creating unfavorable conditions for germination of certain weed species.
- ☆ **Moisture Retention:** By conserving soil moisture, mulches reduce the availability of water for weed seeds, particularly in arid and semi-arid regions.

Types of Mulches for Weed Management

Mulches can be broadly classified into organic and inorganic types, each with distinct characteristics and applications.

1. Organic Mulches

Organic mulches are derived from plant or animal materials and decompose over time, enriching the soil with organic matter. Common examples include:

- ☆ **Crop Residues:** Wheat straw, rice husks, and corn stalks are often used as mulches in field crops.
- ☆ **Wood Chips and Bark:** These are commonly used in orchards and landscapes.
- ☆ **Compost and Manure:** While primarily used for soil fertility, these materials can also suppress weeds when applied as a thick layer.
- ☆ **Grass Clippings and Leaves:** Readily available on farms, these materials are cost-effective for small-scale weed management.

Organic mulches are particularly effective in suppressing annual weeds, as their decomposition releases nutrients and allelopathic compounds. However, their effectiveness depends on the thickness of application and the rate of decomposition.

2. Inorganic Mulches

Inorganic mulches are synthetic or mineral-based materials that do not decompose. They are durable and often used in high-value crops. Examples include:

- **Plastic Mulches:** Black polyethylene films are widely used in horticultural crops to block sunlight and suppress weeds. Colored plastic mulches, such as silver or red, are also used to enhance crop growth while managing weeds. Figure 12.1 indicates more than 95 per cent control of annual weeds of fennel crops. In contrast to black plastic mulch, there are lots of weed in unweeded (control plots)
- **Geotextiles:** These are woven or non-woven fabrics designed for long-term weed suppression in landscapes and orchards.
- **Gravel and Stones:** Commonly used in ornamental gardens, these materials suppress weeds while providing aesthetic value.

Inorganic mulches are particularly effective against perennial weeds due to their long-lasting nature. However, they may require additional management to prevent issues such as soil compaction or heat buildup.

Application Techniques for Effective Weed Control

The effectiveness of mulches in weed management depends on proper application techniques. Key considerations include:

1. **Thickness of Mulch Layer:** A mulch layer of 5–10 cm is generally recommended for effective weed suppression. Thinner layers may allow light penetration, while excessively thick layers can lead to anaerobic soil conditions.
2. **Timing of Application:** Mulches should be applied immediately after planting or before weed emergence to maximize their suppressive effects.
3. **Selection of Mulch Type:** The choice of mulch should align with the crop type, weed species, and environmental conditions. For instance, black plastic mulch is ideal for vegetable crops, while straw mulch is suitable for cereals.
4. **Maintenance:** Organic mulches may need replenishment during the growing season as they decompose. Inorganic mulches should be monitored for damage or displacement.

Flat Sowing (60 DAS)

Ridge Sowing (60 DAS)

Bed sowing (60 DAS)

Figure 12.1. Weed Suppression by Plastic Mulch in Flat, Ridge and Bed Sown Fennel Crop.

Advantages of Mulching in Weed Management

Mulching offers several advantages beyond weed suppression, making it a preferred practice in sustainable agriculture:

- ☆ **Eco-Friendly:** Organic mulches are biodegradable and improve soil health, reducing the need for chemical herbicides.

- **Cost-Effective:** Mulches derived from on-farm residues are inexpensive and readily available.
- **Improved Soil Health:** Organic mulches enhance soil structure, microbial activity, and nutrient availability.
- **Moisture Conservation:** Mulches reduce evaporation, ensuring consistent soil moisture levels for crops.
- **Temperature Regulation:** Mulches protect crops from extreme soil temperatures, promoting uniform growth.

Limitations and Challenges

Despite their benefits, mulches have certain limitations that must be addressed for effective weed management:

- **Labor-Intensive:** Application and maintenance of mulches, especially organic types, can be labor-intensive.
- **Cost of Inorganic Mulches:** Synthetic mulches, such as plastic films, can be expensive and require proper disposal to avoid environmental pollution.
- **Potential for Weed Growth in Mulch Layers:** Organic mulches, if not applied thickly, may harbor weed seeds or allow weeds to grow through gaps.
- **Allelopathy Risks:** While allelopathic compounds can suppress weeds, they may also affect crop growth if not managed carefully.

Recent Advances in Mulching Technology

Innovations in mulching technology have enhanced its effectiveness in weed management:

- **Biodegradable Plastic Mulches:** These mulches decompose naturally, reducing environmental concerns associated with traditional plastic mulches.
- **Sprayable Mulches:** Liquid mulch formulations can be sprayed onto the soil surface, forming a protective layer that suppresses weeds.
- **Integrated Mulching Systems:** Combining mulches with other weed management practices, such as herbicides or cover crops, has shown synergistic effects in controlling weeds.

Case Studies and Research Insights

Recent studies have highlighted the effectiveness of mulches in weed management:

- A study by Kasirajan and Ngouajio (2012) demonstrated that black plastic mulch reduced weed biomass by over 90 per cent in vegetable crops compared to bare soil.

- Research on organic mulches in orchards revealed that wood chips suppressed annual weeds effectively while improving soil organic matter and microbial activity (Chalker-Scott, 2007).

These findings underscore the potential of mulches as a sustainable and versatile tool for weed management in diverse cropping systems.

F. Role of Cover Crops in Weed Suppression

Cover crops are an integral component of modern weed management strategies, offering a sustainable and environmentally friendly approach to suppressing weed growth. By integrating cover crops into cropping systems, farmers can reduce reliance on chemical herbicides, improve soil health, and enhance overall agricultural productivity. This section explores the mechanisms by which cover crops suppress weeds, their benefits, and practical considerations for their use.

Mechanisms of Weed Suppression by Cover Crops

Cover crops suppress weeds through multiple mechanisms, which can be broadly categorized as physical, chemical, and ecological. These mechanisms work synergistically to reduce weed emergence and growth:

- **Physical Suppression**: Cover crops create a dense canopy that shades the soil surface, limiting the amount of sunlight available for weed seed germination and growth. Additionally, the residue left after cover crop termination forms a mulch layer, which acts as a physical barrier to weed emergence. For example, cereal rye (Secale cereale) is known for its ability to produce a thick mulch that effectively suppresses weed growth.
- **Chemical Suppression (Allelopathy)**: Certain cover crops release allelopathic compounds into the soil, which inhibit weed seed germination and growth. For instance, species like rye and sorghum-sudangrass (Sorghum bicolor × Sorghum arundinaceum) produce allelopathic chemicals that target specific weed species. These compounds can reduce the competitiveness of weeds without affecting the main crop.
- **Ecological Suppression**: Cover crops compete with weeds for essential resources such as water, nutrients, and space. By establishing quickly and occupying the niche that weeds would otherwise exploit, cover crops reduce the likelihood of weed establishment. Leguminous cover crops, such as hairy vetch (Vicia villosa), are particularly effective in this regard, as they fix nitrogen while outcompeting weeds. Figure 12.2 indicates that intercropping of cow pea in winter maize not only smoothers the weeds, but also provides nutrition to the crop after the burial of cowpea with earthing up.

Benefits of Using Cover Crops for Weed Management

Incorporating cover crops into cropping systems provides several agronomic and environmental benefits beyond weed suppression. As indicated in Figure 12.2, sowing of cowpea 15 days after sowing spring maize can effectively not only control weeds but the health of the crop is also improved. The cowpea crop was harvested manually and biomass was placed along the crop rows followed by earthing up.

Intercropping of 35 days old Cowpea in 50 Days Old Spring Maize

Incorporation and Earthing Up of Cowpea at 50 DAS Maize

Unweeded Control (50 DAS)

Figure 12.2. Role of Live Mulch on Suppression of Weeds and Growth of Spring Maize.

1. **Reduced Herbicide Dependence**: By suppressing weed populations, cover crops can reduce the need for pre-emergence and post-emergence herbicides. This not only lowers production costs but also minimizes the risk of herbicide resistance in weed populations.
2. **Improved Soil Health**: Cover crops enhance soil structure, increase organic matter content, and improve water infiltration. These soil health benefits indirectly contribute to weed suppression by creating conditions less favorable for weed growth.
3. **Enhanced Biodiversity**: Cover crops promote biodiversity by providing habitat for beneficial insects, soil microorganisms, and other organisms that contribute to natural weed control.
4. **Erosion Control**: The root systems of cover crops stabilize the soil, reducing erosion and runoff. This prevents weed seeds from being transported to new areas and limits the spread of invasive species.

Selection and Management of Cover Crops for Weed Suppression

The effectiveness of cover crops in weed suppression depends on careful selection and management. Key considerations include:

- **Species Selection**: The choice of cover crop species should align with the specific weed challenges, soil type, and climatic conditions of the farm. For example:
  - **Cereal Rye**: Effective for winter cover and weed suppression due to ts rapid growth and high biomass production.
  - **Hairy Vetch**: Ideal for nitrogen fixation and weed competition in spring or summer.
- **Seeding Rate and Timing**: Higher seeding rates and timely planting ensure rapid establishment and canopy closure, which are critical for effective weed suppression. Late planting may allow weeds to establish before the cover crop can compete.
- **Termination Methods**: Proper termination of cover crops is essential to maximize weed suppression while minimizing interference with the main crop. Common methods include mowing, rolling, or herbicide application. For example, rolling-crimping cereal rye at the flowering stage creates a uniform mulch that suppresses weeds effectively.

Challenges and Limitations

While cover crops offer numerous benefits, their adoption is not without challenges. These include:

- **Cost of Establishment**: The initial cost of seeds, planting, and management can be a barrier for some farmers, particularly in small-scale operations.

- ☆ **Potential for Crop Interference**: Poorly managed cover crops can compete with the main crop for resources, leading to yield reductions. Proper timing of termination is critical to avoid this issue.
- ☆ **Weed Adaptation**: Over time, certain weed species may adapt to the presence of cover crops, necessitating the integration of additional weed management strategies.

Case Studies and Research Insights

Recent studies highlight the effectiveness of cover crops in weed suppression. For example, a study by Mirsky *et al.* (2017) demonstrated that cereal rye reduced weed biomass by up to 90 per cent when used as a winter cover crop in no-till systems. Similarly, research by Teasdale *et al.* (2020) found that mixtures of cover crops, such as rye and clover, provided superior weed suppression compared to monocultures, due to their complementary growth habits and resource use.

Chapter 13

Herbicide Resistance in Weeds and their Control Measures

Herbicide resistance in weeds is a major challenge in modern agriculture, reducing the efficacy of chemical weed control and threatening crop productivity. Resistance develops when weed populations evolve mechanisms to survive herbicide applications that previously controlled them. This phenomenon is driven by repeated use of the same herbicide mode of action (MOA) without diversification in weed management strategies (Heap, 2024).

Mechanisms of Herbicide Resistance

Weeds develop resistance through various mechanisms, including:

- ☆ **Target-site resistance (TSR):** Mutations in the herbicide target site reduce herbicide binding, rendering it ineffective. Example: Glyphosate resistance in *Amaranthus palmeri* due to EPSPS gene amplification.
- ☆ **Non-target-site resistance (NTSR):** Enhanced metabolism, reduced herbicide uptake, or sequestration prevents the herbicide from reaching its target. Example: Metabolic resistance to ACCase inhibitors in *Lolium rigidum*.

Factors Contributing to Resistance Development

- ☆ **Repeated use of the same herbicide MOA** without rotation.
- ☆ **Lack of integrated weed management (IWM)** practices.
- ☆ **High weed seed production and dispersal**, leading to rapid spread of resistance genes.

Control Measures for Herbicide-Resistant Weeds

1. **Herbicide Rotation and Mixtures:** Use herbicides with different mode of action to delay resistance development.
2. **Cultural Practices:** Crop rotation, cover cropping, and high planting densities suppress resistant weed populations.
3. **Mechanical Control:** Tillage and hand weeding help manage resistant weed escapes.
4. **Biological Control:** Use of natural weed-suppressing organisms or allelopathic crops.
5. **Precision Weed Management:** Site-specific herbicide application using advanced technologies like drones and AI-based weed detection.

Future Perspectives

Emerging technologies such as RNA interference (RNAi)-based herbicides and gene-editing tools like CRISPR offer potential solutions for managing herbicide-resistant weeds (Duke, 2023).

A. Introduction to Herbicide Resistance

Herbicide resistance is a significant challenge in modern weed management, threatening global agricultural productivity. It occurs when weed populations evolve the ability to survive herbicide applications that previously controlled them. This phenomenon results from repeated use of the same herbicide or herbicide mode of action, leading to the selection of resistant weed biotypes. Understanding herbicide resistance, its mechanisms, and management strategies is crucial for sustainable weed control.

Definition and Importance

Herbicide resistance is defined as the inherited ability of a weed population to survive and reproduce following exposure to a herbicide dose that would normally be lethal to the species (Heap, 2023). This resistance can lead to reduced herbicide efficacy, increased production costs, and lower crop yields. As resistant weed populations spread, farmers must adopt integrated weed management strategies to mitigate resistance development.

Mechanisms of Herbicide Resistance

Weeds develop resistance through various mechanisms, which can be classified into two major categories:

1 Target-Site Resistance (TSR)

This occurs when mutations in the target site of the herbicide prevent the herbicide from binding effectively. As a result, the herbicide fails to inhibit the intended biochemical pathway. Examples include:

- **Acetolactate Synthase (ALS) Inhibitor Resistance:** Mutations in the ALS gene prevent herbicides such as imazapyr and sulfonylureas from binding, rendering them ineffective.
- **Acetyl-CoA Carboxylase (ACCase) Inhibitor Resistance:** Mutations in the ACCase enzyme confer resistance to herbicides like clethodim and sethoxydim, commonly used for grass weed control.

2. Non-Target-Site Resistance (NTSR)

This form of resistance involves mechanisms that reduce the herbicide's effectiveness without altering the target site. These mechanisms include:

- **Enhanced Metabolism:** Resistant weeds break down herbicides more rapidly before they can exert their toxic effects. This is common in resistance to atrazine and glyphosate.
- **Reduced Herbicide Absorption or Translocation:** Some resistant weeds limit herbicide uptake or movement within the plant, reducing its efficacy.
- **Sequestration:** Certain weeds can compartmentalize herbicides into vacuoles or other cellular structures, preventing them from reaching their target site.

Factors Contributing to Herbicide Resistance

Several factors accelerate the development of herbicide resistance in weed populations:

- **Repeated Use of the Same Herbicide Mode of Action (MOA):** Continuous application of herbicides with the same MOA exerts strong selection pressure, favoring resistant biotypes.
- **Monoculture and Lack of Crop Rotation:** Growing the same crop repeatedly without rotation encourages the dominance of specific weed species and resistance development.
- **Inadequate Herbicide Application:** Sub-lethal herbicide doses can allow partially resistant individuals to survive and reproduce, increasing resistance levels over time.
- **Limited Use of Integrated Weed Management (IWM):** Over-reliance on chemical control without incorporating cultural, mechanical, and biological methods accelerates resistance evolution.

Examples of Herbicide-Resistant Weeds

Several weed species have evolved resistance to multiple herbicide modes of action, posing significant challenges to weed control. Some notable examples include:

- ✰ ***Amaranthus palmeri* (Palmer Amaranth):** Resistant to glyphosate and ALS inhibitors, making it one of the most problematic weeds in North America.
- ✰ ***Lolium rigidum* (Rigid Ryegrass):** Exhibits resistance to multiple herbicide classes, including ACCase inhibitors and glyphosate, particularly in Australia.
- ✰ ***Conyza canadensis* (Horseweed):** Known for glyphosate resistance, making it difficult to control in no-till farming systems.

Herbicide Resistance Management Strategies

To mitigate herbicide resistance, farmers and agronomists must adopt proactive management strategies, including:

- ✰ **Herbicide Rotation and Mixtures:** Using herbicides with different modes of action in rotation or tank mixtures reduces selection pressure on weed populations.
- ✰ **Cultural Practices:** Crop rotation, cover cropping, and competitive crop planting help suppress weed growth and reduce reliance on herbicides.
- ✰ **Mechanical Control:** Tillage, mowing, and hand-weeding can physically remove resistant weed populations.
- ✰ **Use of Precision Agriculture:** Technologies such as site-specific herbicide application and weed mapping improve herbicide efficiency and reduce overuse.
- ✰ **Integrated Weed Management (IWM):** Combining chemical, cultural, mechanical, and biological control methods ensures long-term weed suppression and delays resistance development.

Herbicide resistance continues to be a major concern in global agriculture, necessitating a shift towards sustainable weed management practices. By understanding the mechanisms and contributing factors, agricultural professionals can implement effective resistance management strategies to maintain herbicide efficacy and protect crop productivity.

B. Mechanisms of Herbicide Resistance

Herbicide resistance is a major challenge in modern weed management. It occurs when weed populations evolve the ability to survive herbicide applications that previously controlled them. This resistance develops due to repeated use of the same herbicide or herbicide mode of action, leading to the selection of resistant biotypes. Understanding the mechanisms of herbicide resistance is crucial for developing effective weed management strategies.

Types of Herbicide Resistance

Herbicide resistance in weeds can be classified into two main categories:

1. **Target-Site Resistance (TSR)**
2. **Non-Target-Site Resistance (NTSR)**

Each type involves different physiological and biochemical changes in the weed, allowing it to survive herbicide applications.

1. Target-Site Resistance (TSR)

Target-site resistance occurs when mutations or alterations in the herbicide's target site prevent the herbicide from binding effectively. This is the most common mechanism of resistance and is often caused by point mutations in specific genes.

Mechanisms of Target-Site Resistance

- **Amino Acid Substitutions:** A single nucleotide change in the gene encoding the herbicide target enzyme can lead to an amino acid substitution, reducing herbicide binding. For example:
- Glyphosate resistance in *Amaranthus palmeri* is caused by mutations in the **5-enolpyruvylshikimate-3-phosphate synthase (EPSPS)** enzyme (Sammons and Gaines, 2014).
- Acetolactate synthase (ALS) inhibitor resistance in *Lolium rigidum* is due to mutations in the **ALS gene**.
- **Gene Amplification:** Some resistant weeds produce multiple copies of the target-site gene, increasing enzyme production and reducing herbicide effectiveness.
- Glyphosate-resistant *Amaranthus palmeri* has been found to contain multiple copies of the **EPSPS gene**, leading to overproduction of the enzyme (Gaines *et al.*, 2010).
- **Structural Changes in Target Enzymes:** Some resistant weeds modify the structure of the target enzyme, reducing herbicide binding without affecting normal enzyme function.

2. Non-Target-Site Resistance (NTSR)

Non-target-site resistance involves mechanisms that reduce the herbicide's ability to reach or affect its target site. This type of resistance is often more complex and can confer resistance to multiple herbicide groups.

Mechanisms of Non-Target-Site Resistance

- **Enhanced Metabolism:** Resistant weeds break down herbicides more rapidly before they reach toxic levels. This is often due to increased activity of detoxification enzymes such as:

- **Cytochrome P450 monooxygenases**
- **Glutathione S-transferases (GSTs)**
- **Glycosyltransferases**

For example, *Alopecurus myosuroides* (blackgrass) has evolved resistance to ACCase and ALS inhibitors through enhanced metabolism.

- **Reduced Herbicide Uptake or Translocation:** Some resistant weeds limit herbicide absorption or movement within the plant.
  - Glyphosate-resistant *Conyza canadensis* exhibits reduced herbicide translocation, preventing it from reaching lethal concentrations in meristematic tissues.
- **Sequestration or Compartmentalization:** Some plants isolate herbicides in vacuoles or other cellular compartments, preventing them from reaching their target site.
  - Glyphosate-resistant *Lolium spp.* have been shown to sequester glyphosate into vacuoles, reducing its effectiveness.

Cross-Resistance and Multiple Resistance

- **Cross-Resistance:** A single resistance mechanism provides resistance to multiple herbicides within the same mode of action.

  Example: ALS-resistant *Amaranthus tuberculatus* is resistant to multiple ALS-inhibiting herbicides due to a single target-site mutation.
- **Multiple Resistance:** A weed population evolves resistance to two or more herbicide modes of action through different mechanisms.

  Example: *Lolium rigidum* populations in Australia have developed resistance to glyphosate, ALS inhibitors, and ACCase inhibitors through both TSR and NTSR mechanisms.

Implications for Weed Management

Understanding herbicide resistance mechanisms is essential for developing integrated weed management strategies. The widespread occurrence of both TSR and NTSR highlights the need for diversified control methods, including herbicide rotation, cultural practices, and non-chemical approaches.

C. Common Herbicide-Resistant Weed Species

Herbicide resistance in weeds is a growing challenge in modern agriculture, threatening crop productivity and sustainability. Resistance develops when weed populations are repeatedly exposed to the same herbicide mode of action, leading to the survival and proliferation of resistant biotypes. Understanding the most problematic herbicide-resistant weed species is crucial for developing effective management strategies. This section discusses some of the most widespread and economically significant herbicide-resistant weed species globally.

1. *Amaranthus* spp. (Pigweeds)

Several species within the Amaranthus genus have developed resistance to multiple herbicide modes of action, making them some of the most troublesome weeds in row crop agriculture.

a) Palmer Amaranth (*Amaranthus palmeri*)

- ☆ One of the most aggressive and competitive weed species in North America.
- ☆ Resistant to glyphosate (EPSP synthase inhibitor), ALS inhibitors, and PPO inhibitors.
- ☆ Can produce over 500,000 seeds per plant, ensuring rapid spread and persistence.
- ☆ Commonly found in soybean, cotton, and corn fields.

b) Waterhemp (*Amaranthus tuberculatus*)

- ☆ Highly adaptable and capable of evolving resistance quickly.
- ☆ Documented resistance to glyphosate, ALS inhibitors, PPO inhibitors, and HPPD inhibitors.
- ☆ Dioecious nature (separate male and female plants) promotes genetic diversity, accelerating resistance development.
- ☆ Major weed in the Midwest United States, particularly in corn and soybean production systems.

2. *Lolium* spp. (Ryegrass)

Ryegrass species are notorious for their ability to develop herbicide resistance, particularly in cereal cropping systems.

a) Italian Ryegrass (*Lolium multiflorum*)

- ☆ Widespread resistance to glyphosate, ACCase inhibitors, and ALS inhibitors.
- ☆ Common in wheat, orchards, and vineyards, particularly in Europe, North America, and Australia.
- ☆ High seed production and dormancy contribute to its persistence in cropping systems.

b Perennial Ryegrass (*Lolium perenne*)

- ☆ Increasing resistance to glyphosate and other herbicide groups.
- ☆ Found in pastures, turfgrass systems, and cereal crops.
- ☆ Difficult to control due to its ability to regenerate from tillers.

c) Rigid Ryegrass (*Lolium rigidum*)

- ☆ One of the most herbicide-resistant weed species globally.
- ☆ Found in wheat and barley fields in Australia and the Mediterranean region.
- ☆ Resistance to glyphosate, ACCase inhibitors, ALS inhibitors, and paraquat.

3. *Phalaris* spp. (Canary Grass)

Canary grass species are major herbicide-resistant weeds in cereal cropping systems, especially wheat.

a) Little Seed Canary Grass (*Phalaris minor*)

- ☆ Resistance to ACCase inhibitors (clodinafop, fenoxaprop), ALS inhibitors (sulfosulfuron, metsulfuron), and cases of multiple resistance reported.
- ☆ Common in wheat and barley fields, especially in North India, Pakistan, and parts of Europe.
- ☆ High seed production, prolonged emergence period, and close resemblance to wheat seedlings make it challenging to control.

b) Hood Canary Grass (*Phalaris paradoxa*)

- ☆ Reported resistance to ACCase-inhibiting herbicides in Mediterranean regions and parts of Australia.
- ☆ Thrives in winter cereals and fallow lands with heavy infestations.
- ☆ Competitive growth, rapid tillering, and extended seed dormancy contribute to its persistence.

4. *Echinochloa* spp. (Barnyardgrass)

Barnyardgrass species are among the most problematic weeds in rice production worldwide.

a) Barnyardgrass (*Echinochloa crus-galli*)

- ☆ Resistance to ALS inhibitors, ACCase inhibitors, and quinclorac (synthetic auxins).
- ☆ Common in flooded and dry-seeded rice systems.
- ☆ High seed production and ability to mimic rice plants make it difficult to manage.

b) Junglerice (*Echinochloa colona*)

- ☆ Increasing resistance to glyphosate in no-till and direct-seeded rice systems.

- ✰ Found in tropical and subtropical regions, particularly in Asia and South America.
- ✰ Rapid emergence and growth make it highly competitive with crops.

5. Horseweed (*Conyza canadensis*)

- ✰ One of the first weeds to develop glyphosate resistance in North America.
- ✰ Also resistant to ALS inhibitors and paraquat.
- ✰ A major problem in no-till cropping systems, where it emerges before planting.
- ✰ Produces wind-dispersed seeds, allowing rapid spread across large areas.

6. *Eleusine indica* (Goosegrass)

- ✰ Resistance to glyphosate, ACCase inhibitors, and dinitroaniline herbicides.
- ✰ Common in tropical and subtropical regions, particularly in turfgrass, roadsides, and row crops.
- ✰ Strong tillering ability and seed dormancy contribute to its persistence.

7. *Avena fatua* (Wild Oat)

- ✰ Resistance to ACCase inhibitors, ALS inhibitors, and triallate (thiocarbamate herbicide).
- ✰ A major weed in wheat and barley fields in North America, Europe, and Australia.
- ✰ High seed dormancy and prolonged emergence period make it difficult to control.

8. *Sorghum halepense* (Johnsongrass)

- ✰ Resistance to glyphosate and ALS inhibitors.
- ✰ A highly invasive perennial weed found in corn, soybean, and sorghum fields.
- ✰ Spreads through seeds and rhizomes, making mechanical control challenging.

9. *Digitaria* spp. (Crabgrasses)

- ✰ Resistance to ACCase inhibitors and dinitroaniline herbicides.
- ✰ Common in turfgrass, pastures, and row crops.
- ✰ High seed production and rapid growth contribute to its persistence.

10. Kochia (*Kochia scoparia*)

- Resistance to glyphosate, ALS inhibitors, and dicamba (synthetic auxin).
- A major weed in dryland cropping systems in North America.
- Wind-dispersed seeds allow rapid colonization of new areas.

Figure 13.1 illustrates the number of herbicide-resistant species across the top 10 crops globally. Wheat has the highest number of resistant species at 86, making it the most affected crop. Corn (maize) follows with 66 resistant species, showing significant herbicide resistance issues. Rice and soybean also show high resistance cases, with 57 and 53 species respectively.

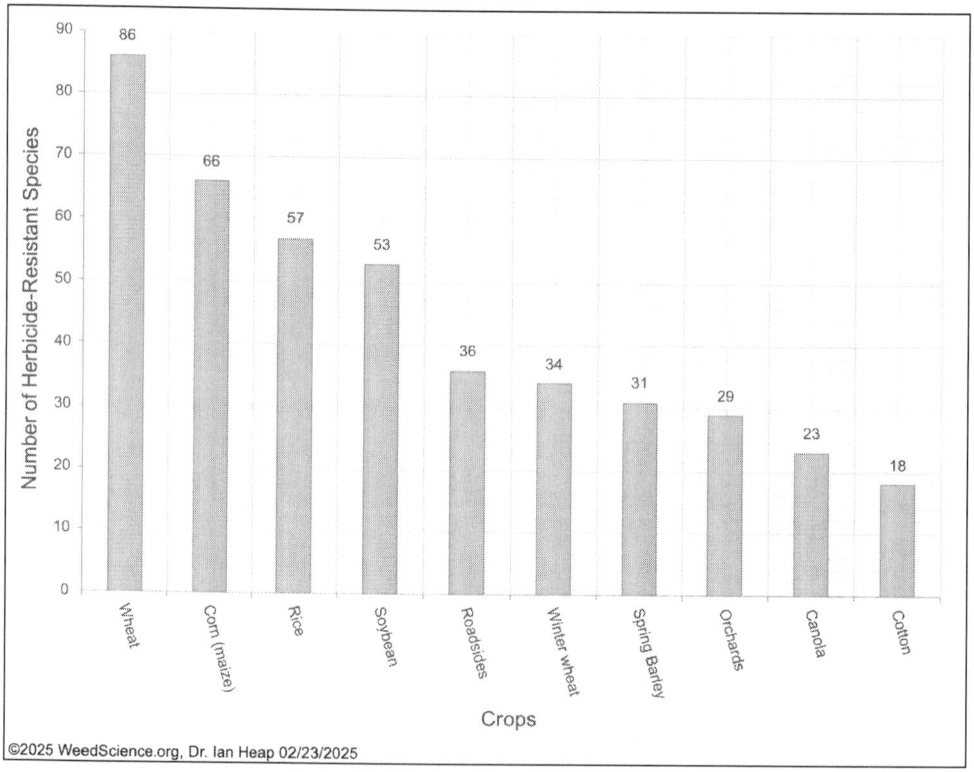

Figure 13.1. Number of Herbicide-Resistant Species by Crop (top 10).

Roadsides, though not a crop, have 36 resistant species, indicating herbicide pressure beyond agriculture. Winter wheat and spring barley have 34 and 31 resistant species, highlighting challenges in cereal production systems. Orchards also face resistance issues with 29 herbicide-resistant species. Canola and cotton have relatively lower numbers at 23 and 18 resistant species, respectively, but still require monitoring.

The figure suggests that major cereal and grain crops are more prone to herbicide resistance.

This emphasizes the need for diverse weed management practices in these crop systems to reduce reliance on herbicides.

Management Implications

The widespread occurrence of herbicide-resistant weeds necessitates integrated weed management (IWM) strategies, including:

- ☆ **Herbicide rotation and mixtures** to delay resistance development.
- ☆ **Cultural practices** such as crop rotation, cover cropping, and competitive crop varieties.
- ☆ **Mechanical control** including tillage and mowing where feasible.
- ☆ **Biological control** using natural weed-suppressing organisms.

Understanding the biology and resistance mechanisms of these weed species is essential for designing effective, sustainable weed management programs.

D. Factors Contributing to Herbicide Resistance

Herbicide resistance is a significant challenge in modern weed management, leading to reduced efficacy of chemical control methods and increased production costs. Resistance develops when weed populations evolve mechanisms to survive herbicide applications that previously controlled them. Several factors contribute to the development and spread of herbicide resistance, including biological, genetic, and management-related aspects. Understanding these factors is crucial for designing effective resistance management strategies.

1. Repeated Use of the Same Herbicide Mode of Action (MOA)

One of the primary drivers of herbicide resistance is the continuous use of herbicides with the same mode of action. When the same herbicide or herbicide group is applied repeatedly, susceptible weed populations are eliminated, while resistant individuals survive and reproduce. Over time, this leads to a shift in the weed population, favoring resistant biotypes.

- ☆ Example: Glyphosate resistance in *Amaranthus palmeri* (Palmer amaranth) has emerged due to the extensive use of glyphosate-based herbicides in glyphosate-resistant cropping systems (Norsworthy *et al.*, 2012).
- ☆ Scientific Principle: Selection pressure increases when a single MOA is used continuously, accelerating the evolution of resistance.

2. Lack of Herbicide Rotation and Tank Mixing

Failure to rotate herbicides with different MOAs or use tank mixtures with multiple effective MOAs contributes to resistance development. When a single herbicide is used repeatedly, weeds with resistance traits survive, leading to a resistant population over time.

- ☆ Best Practice: Using herbicide mixtures that contain multiple MOAs can delay resistance development by targeting weeds through different biochemical pathways.
- ☆ Example: Studies have shown that tank-mixing herbicides with different MOAs reduces the likelihood of resistance compared to sequential applications of single MOA herbicides (Beckie and Harker, 2017).

3. Weed Biology and Reproductive Traits

Certain weed species have biological characteristics that make them more prone to developing resistance. These include:

- ☆ **High Seed Production**: Weeds that produce a large number of seeds can rapidly spread resistance traits.
- ☆ **Short Life Cycle**: Annual weeds with multiple generations per year evolve resistance faster than perennials.
- ☆ **Cross-Pollination**: Weed species capable of cross-pollination can spread resistance genes more quickly than self-pollinating species.
- ☆ Example: *Lolium rigidum* (annual ryegrass) is highly prone to herbicide resistance due to its high genetic variability and cross-pollination ability (Busi *et al.*, 2013).

4. Sub-Lethal Herbicide Doses

Applying herbicides at doses lower than the recommended rate can contribute to resistance by allowing partially resistant individuals to survive and reproduce. Sub-lethal doses do not completely eliminate susceptible weeds, leading to the selection of individuals with slight resistance, which can accumulate over time.

- ☆ Example: Research has shown that repeated exposure to sub-lethal doses of glyphosate can lead to resistance in *Echinochloa colona* (junglerice) populations (Fernández-Moreno *et al.*, 2017).
- ☆ Best Practice: Always apply herbicides at the recommended label rates to ensure effective weed control and minimize resistance selection.

5. Lack of Crop Rotation and Diversified Weed Management

Monoculture cropping systems that rely heavily on herbicides for weed control create an environment conducive to resistance development. Crop rotation and integrated weed management (IWM) strategies can help mitigate this risk.

- **Impact of Monoculture**: Continuous cultivation of the same crop allows specific weed species to dominate and adapt to herbicide applications.
- **Benefits of Crop Rotation**: Rotating crops with different growth habits and planting times disrupts weed life cycles and reduces reliance on specific herbicides.
- Example: Rotating soybean with wheat and using different herbicide programs has been shown to reduce glyphosate-resistant *Conyza canadensis* (horseweed) populations (Owen *et al.*, 2015).

6. Herbicide Persistence and Residual Activity

Some herbicides have long residual activity in the soil, which can contribute to resistance if weeds are exposed to sub-lethal doses over extended periods. Additionally, herbicides with prolonged soil persistence may select for resistant weed populations over multiple growing seasons.

- Example: Atrazine-resistant *Chenopodium album* (common lambsquarters) has been reported in fields with repeated atrazine applications over several years (Heap, 2024).
- Best Practice: Use herbicides with different persistence levels and incorporate non-chemical weed control methods to reduce selection pressure.

7. Lack of Proper Weed Scouting and Resistance Monitoring

Failure to monitor weed populations for early signs of resistance can lead to widespread resistance before corrective measures are implemented. Regular scouting and resistance testing help detect resistant biotypes early, allowing for timely management adjustments.

- **Key Indicators of Resistance**:
- Poor weed control despite correct herbicide application.
- Surviving weeds of the same species after herbicide treatment.
- Gradual increase in resistant weed populations over time.
- Best Practice: Conduct field scouting before and after herbicide applications and implement resistance testing when control failures occur.

8. Limited Adoption of Integrated Weed Management (IWM) Strategies

Relying solely on herbicides for weed control increases the risk of resistance. Integrated Weed Management (IWM) combines chemical, cultural, mechanical, and biological control methods to reduce selection pressure on herbicides.

Key IWM Strategies

- ✰ Use of cover crops to suppress weed emergence.
- ✰ Mechanical weed control methods such as tillage and mowing.
- ✰ Biological control using natural weed-suppressing organisms.
- ✰ Example: Studies have shown that integrating cover crops with herbicide programs reduces herbicide-resistant *Amaranthus* species populations (Mirsky *et al.*, 2013).

9. Gene Flow and Resistance Spread

Herbicide resistance can spread between weed populations through gene flow mechanisms such as pollen movement, seed dispersal, and human activities. Resistant weed seeds can be transported by farm machinery, irrigation water, or wind, leading to the establishment of resistant populations in new areas.

- ✰ Example: Glyphosate-resistant *Ambrosia artemisiifolia* (common ragweed) has spread across multiple regions due to pollen-mediated gene flow (VanGessel, 2001).
- ✰ Best Practice: Implement sanitation measures such as cleaning equipment and using certified weed-free seeds to prevent resistance spread.

E. Detection and Diagnosis of Herbicide Resistance

Herbicide resistance is a significant challenge in modern weed management, leading to reduced efficacy of chemical control methods and increased production costs. Detecting and diagnosing herbicide resistance early is crucial for implementing effective management strategies. This section discusses the methods used to identify herbicide resistance, including field observations, bioassays, biochemical tests, and molecular techniques.

Field-Based Detection of Herbicide Resistance

Field observations are the first step in identifying potential herbicide resistance. Certain signs indicate that weeds may have developed resistance to a specific herbicide:

- ✰ **Survival of treated weeds**: If a herbicide is applied at the recommended rate and timing, but certain weed populations survive while others die, resistance is suspected.
- ✰ **Patchy weed control**: Resistant weeds often appear in clusters, indicating that susceptible individuals have been eliminated while resistant ones persist.
- ✰ **Repeated failure of herbicide applications**: If a herbicide that was previously effective no longer controls the target weed species despite proper application, resistance may be developing.

- **Multiple resistance**: If weeds survive applications of herbicides from different modes of action, they may have evolved multiple resistance mechanisms.

Bioassays for Herbicide Resistance Diagnosis

Bioassays are laboratory or greenhouse tests that confirm herbicide resistance by comparing the response of suspected resistant weed populations to known susceptible populations. These tests include:

- **Whole-plant bioassays**: Seeds or seedlings of suspected resistant weeds are grown and treated with different herbicide doses. Resistance is confirmed if the plants survive doses that normally kill susceptible populations.
- **Seed germination and seedling growth assays**: Seeds are germinated in petri dishes or soil with herbicide-treated media to assess resistance levels.
- **Hydroponic bioassays**: Weed seedlings are grown in nutrient solutions containing herbicides to evaluate resistance.

Bioassays provide a practical and cost-effective method for detecting resistance, but they require several weeks to obtain results.

Biochemical Tests for Herbicide Resistance

Biochemical assays detect specific resistance mechanisms at the enzyme level. These tests are particularly useful for identifying target-site resistance and metabolic resistance.

- **Acetolactate synthase (ALS) enzyme assays**: Used to detect resistance to ALS-inhibiting herbicides (*e.g.*, sulfonylureas, imidazolinones). Resistant weeds exhibit reduced sensitivity to ALS inhibitors.
- **Acetyl-CoA carboxylase (ACCase) enzyme assays**: Used for detecting resistance to ACCase-inhibiting herbicides (*e.g.*, aryloxyphenoxypropionates, cyclohexanediones).
- **Glutathione S-transferase (GST) assays**: Identify metabolic resistance by measuring the activity of detoxifying enzymes that break down herbicides before they reach their target sites.

Biochemical tests provide faster results than bioassays but require specialized laboratory equipment and expertise.

Molecular Techniques for Herbicide Resistance Detection

Advancements in molecular biology have enabled precise and rapid detection of herbicide resistance at the genetic level. These techniques identify specific mutations or gene expressions associated with resistance.

- ☆ **Polymerase Chain Reaction (PCR) assays**: Detect point mutations in genes encoding herbicide target enzymes, such as ALS, ACCase, and EPSPS (5-enolpyruvylshikimate-3-phosphate synthase, associated with glyphosate resistance) (Heap, 2023).
- ☆ **Quantitative PCR (qPCR)**: Measures gene expression levels to detect enhanced metabolism resistance, where resistant weeds produce higher amounts of detoxifying enzymes.
- ☆ **Next-Generation Sequencing (NGS)**: Provides a comprehensive analysis of resistance genes, enabling the discovery of novel resistance mechanisms.
- ☆ **Genotyping-by-Sequencing (GBS)**: Identifies genetic markers linked to herbicide resistance, allowing for early detection in weed populations.

Molecular techniques offer high specificity and accuracy but require advanced laboratory facilities and expertise.

Herbicide Resistance Screening in Weed Populations

To confirm herbicide resistance, researchers and agronomists conduct systematic screening of weed populations using a combination of the above methods. The process includes:

1. **Field sampling**: Collecting seeds or plant samples from suspected resistant populations.
2. **Greenhouse or laboratory testing**: Conducting bioassays, biochemical tests, or molecular analyses.
3. **Data interpretation**: Comparing results with known susceptible populations to determine resistance status.
4. **Resistance mapping**: Documenting resistant populations to guide management decisions.

Key Challenges in Herbicide Resistance Detection

Despite advancements in detection methods, several challenges remain:

- ☆ **Delayed diagnosis**: Resistance is often detected only after herbicide failure, making management more difficult.
- ☆ **Complex resistance mechanisms**: Some weeds exhibit multiple resistance mechanisms, complicating detection and control.
- ☆ **Limited access to molecular tools**: Advanced molecular techniques are not widely available in all regions due to cost and technical constraints.
- ☆ **Genetic variability**: Resistance evolution varies among weed species and populations, requiring continuous monitoring.

Conclusion

Detecting and diagnosing herbicide resistance is essential for effective weed management. A combination of field observations, bioassays, biochemical tests, and molecular techniques provides a comprehensive approach to identifying resistance. Early detection allows for timely intervention, reducing the spread of resistant weed populations and preserving the efficacy of herbicides.

F. Global Status of Herbicide Resistance in Weeds

Herbicide resistance in weeds has become a significant challenge in modern agriculture, threatening global food security and sustainable weed management. The widespread and repeated use of herbicides has led to the evolution of resistant weed populations, reducing the efficacy of chemical control methods. This section explores the current global status of herbicide resistance, highlighting key resistant weed species, affected regions, and the implications for weed management strategies.

Due to continuous use of herbicides, especially with the same mode of action, the resistance in weed is increasing at a very fast rate all over the world (Figure 13.2). The figure shows a steady increase in the number of unique herbicide-resistant weed cases globally from 1950 to 2025. Resistance cases remained negligible until around 1980, after which a sharp upward trend began. The growth accelerated significantly between 1990 and 2015, indicating increased herbicide use and selection pressure. After 2015, the growth rate appears to slow slightly, approaching a plateau near 525 cases by 2025. This trend highlights the urgent need for integrated weed management strategies worldwide.

Understanding Herbicide Resistance

Herbicide resistance is the inherited ability of a weed population to survive and reproduce despite the application of a herbicide that previously controlled it. Resistance develops due to natural selection, where repeated exposure to the same herbicide exerts selection pressure, favoring resistant biotypes.

There are two primary mechanisms of herbicide resistance:

- ☆ **Target-site resistance (TSR):** Mutations in the target enzyme reduce herbicide binding, rendering it ineffective.
- ☆ **Non-target-site resistance (NTSR):** Enhanced metabolism, reduced herbicide absorption, or sequestration prevents the herbicide from reaching its target site.

Global Trends in Herbicide Resistance

Herbicide resistance has been reported in 266 weed species across 71 countries, affecting 21 of the 31 known herbicide sites of action (Heap, 2024).

The problem is most severe in regions with intensive herbicide use, such as North America, South America, Europe, and Australia.

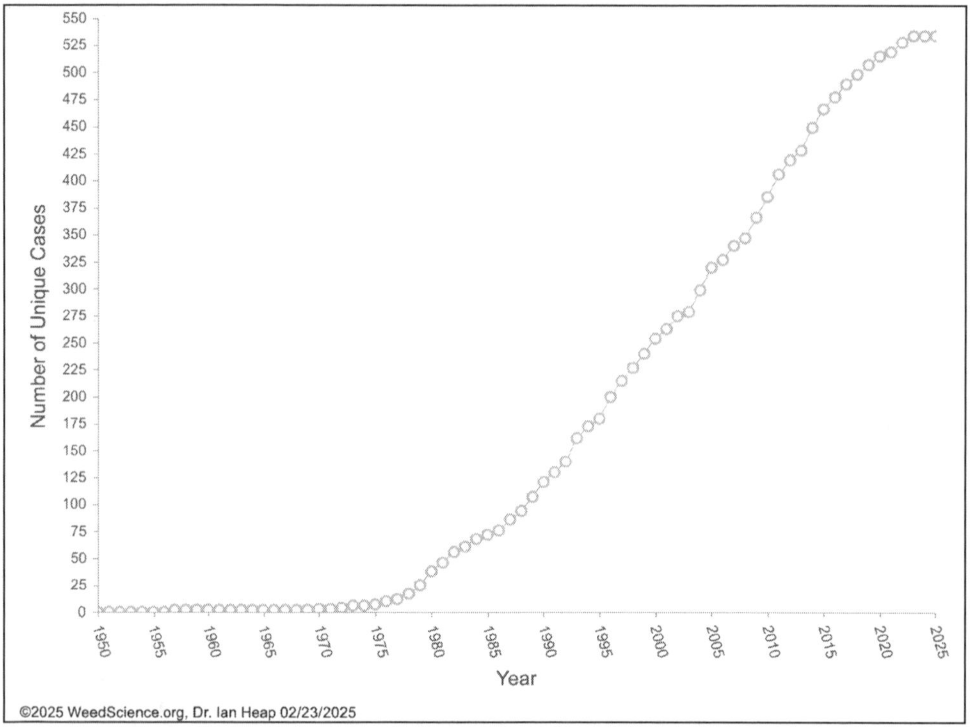

Figure 13.1: Chronological Increase in Resistant Weeds Globally.

1. North America

- ✰ The United States has the highest number of herbicide-resistant weed cases, with resistance confirmed in over 160 weed species.
- ✰ Glyphosate-resistant *Amaranthus palmeri* (Palmer amaranth) and *Amaranthus tuberculatus* (waterhemp) pose severe challenges in soybean, cotton, and corn production.
- ✰ Multiple resistance to glyphosate and other herbicides (*e.g.*, ALS inhibitors, PPO inhibitors) is widespread, complicating control efforts.

2. South America

- ✰ Brazil and Argentina face severe resistance issues due to extensive soybean and maize cultivation with glyphosate-resistant crops.
- ✰ *Echinochloa crus-galli* (barnyardgrass) and *Lolium* spp. (ryegrass) have developed resistance to multiple herbicide modes of action.
- ✰ Resistance to ACCase inhibitors and ALS inhibitors is increasing, reducing the effectiveness of post-emergence herbicides.

3. Europe

- Resistance in *Alopecurus myosuroides* (blackgrass) and *Lolium rigidum* (rigid ryegrass) is a major concern in wheat production.
- The United Kingdom, France, and Germany report widespread resistance to ALS inhibitors and ACCase inhibitors.
- Non-target-site resistance is particularly common, making chemical control less reliable.

4. Australia

- ***Lolium rigidum*** (annual ryegrass) is the most problematic resistant weed, with resistance to multiple herbicide groups, including glyphosate, ALS inhibitors, and ACCase inhibitors.
- Resistance evolution is driven by continuous cropping systems and reliance on herbicides as the primary weed control method.
- Integrated weed management (IWM) strategies, such as harvest weed seed control, are being adopted to mitigate resistance.

5. Asia and Africa

- In Asia, resistance in *Echinochloa* spp. (barnyardgrass) and *Leptochloa chinensis* (Chinese sprangletop) threatens rice production, particularly in India, China, and Southeast Asia.
- Africa has fewer documented cases, but resistance in *Rottboellia cochinchinensis* (itchgrass) and *Eleusine indica* (goosegrass) is emerging in maize and sugarcane fields.

Herbicide Resistance in Major Herbicide Groups

The most affected herbicide groups include:

- **Glyphosate (EPSPS inhibitors):** Resistance is widespread in *Amaranthus* spp., *Conyza* spp., and *Lolium* spp.
- **ALS inhibitors (Acetolactate synthase inhibitors):** Resistance is common in *Alopecurus myosuroides, Amaranthus spp.,* and *Echinochloa* spp.
- **ACCase inhibitors (Acetyl-CoA carboxylase inhibitors):** Resistance is prevalent in *Lolium* spp. and *Avena fatua* (wild oat).
- **PPO inhibitors (Protoporphyrinogen oxidase inhibitors):** Resistance is increasing in *Amaranthus* spp., particularly in the United States.

Implications for Weed Management

Herbicide resistance has serious consequences for agricultural productivity and sustainability:

- ☆ **Reduced herbicide efficacy:** Farmers must use higher doses or alternative herbicides, increasing production costs.
- ☆ **Limited chemical control options:** Multiple resistance makes some weed populations nearly impossible to control with herbicides alone.
- ☆ **Increased reliance on non-chemical methods:** Mechanical, cultural, and biological control methods are becoming essential components of integrated weed management.
- ☆ **Environmental and economic impact:** Overuse of herbicides can lead to soil and water contamination, while resistance management increases input costs for farmers.

Future Outlook and Research Directions

To combat herbicide resistance, researchers and agronomists are focusing on:

- ☆ **Developing new herbicide modes of action** to target resistant weed populations.
- ☆ **Enhancing integrated weed management (IWM)** by combining chemical, mechanical, and cultural control strategies.
- ☆ **Using precision agriculture technologies** such as site-specific herbicide application and robotic weed control.
- ☆ **Exploring genetic and biotechnological solutions** like RNA interference (RNAi) and gene editing for targeted weed suppression.

Herbicide resistance remains a dynamic and evolving challenge, requiring continuous monitoring and adaptive management strategies to sustain global agricultural productivity.

G. Integrated Weed Management (IWM) Strategies

Integrated Weed Management (IWM) is a sustainable approach to weed control that combines multiple methods to reduce weed populations while minimizing environmental impact. Unlike sole reliance on chemical herbicides, IWM integrates cultural, mechanical, biological, and chemical control methods to achieve long-term weed suppression. This approach enhances crop productivity, reduces herbicide resistance, and promotes ecological balance in agricultural systems.

Principles of Integrated Weed Management

IWM is based on the following key principles:

- ☆ **Diversity in Control Methods**: Utilizing multiple weed control techniques to prevent adaptation and resistance.
- ☆ **Prevention and Early Intervention**: Implementing measures to stop weed establishment and spread.

- **Sustainable Practices**: Reducing reliance on herbicides to minimize environmental and health risks.
- **Economic Viability**: Ensuring that weed control measures are cost-effective and practical for farmers.

Components of Integrated Weed Management

1. Cultural Control

Cultural practices enhance crop competitiveness against weeds by modifying the growing environment. These include:

- **Crop Rotation**: Alternating crops with different growth habits disrupts weed life cycles. For example, rotating cereals with legumes can reduce the prevalence of specific weed species (Chauhan and Mahajan, 2014).
- **Cover Crops**: Planting cover crops like clover or rye suppresses weed growth by shading and outcompeting weeds for nutrients.
- **Optimum Planting Density**: Higher crop density reduces available space and resources for weed establishment.
- **Mulching**: Organic or synthetic mulches block sunlight, preventing weed seed germination.

2. Mechanical and Physical Control

These methods involve physical removal or destruction of weeds and include:

- **Tillage**: Plowing or harrowing disrupts weed roots and buries weed seeds. However, excessive tillage can lead to soil erosion.
- **Hand Weeding and Hoeing**: Effective for small-scale farming but labor-intensive.
- **Flaming and Thermal Weeding**: Using heat to destroy weed tissues, often applied in organic farming.
- **Mowing and Cutting**: Reduces seed production in perennial weeds like Canada thistle (*Cirsium arvense*).

3. Biological Control

Biological weed control involves using natural enemies such as insects, pathogens, or competitive plants to suppress weed populations. Examples include:

- **Insect Herbivores**: Certain insects, like the *Chrysolina* beetle, effectively control invasive weeds such as St. John's wort (*Hypericum perforatum*).
- **Pathogens**: Fungal pathogens like *Colletotrichum gloeosporioides* have been explored for biocontrol of weeds like sicklepod (*Senna obtusifolia*).

- **Allelopathic Crops**: Some crops, such as rye (*Secale cereale*), release natural chemicals that inhibit weed germination.

4. Chemical Control

Herbicides remain an essential component of IWM but should be used judiciously to prevent resistance. Strategies include:

- **Herbicide Rotation and Mixtures**: Using different modes of action reduces the risk of herbicide-resistant weed populations.
- **Site-Specific Application**: Precision spraying using GPS and sensor technologies minimizes herbicide use while targeting weed-infested areas.
- **Pre-emergence and Post-emergence Herbicides**: Pre-emergence herbicides prevent weed seed germination, while post-emergence herbicides control actively growing weeds.

5. Preventive Measures

Preventing weed introduction and spread is crucial in IWM. Key measures include:

- **Certified Weed-Free Seeds**: Using clean seed material prevents the introduction of weed seeds.
- **Sanitation of Farm Equipment**: Cleaning machinery before moving between fields reduces weed seed dispersal.
- **Border Management**: Controlling weeds along field edges and irrigation channels prevents their spread into cultivated areas.

Emerging Technologies in IWM

Advancements in agricultural technology are enhancing IWM effectiveness. Some notable innovations include:

- **Precision Agriculture**: Drones and AI-powered weed detection systems enable targeted herbicide application, reducing chemical use and costs.
- **Autonomous Weeding Robots**: Machines equipped with cameras and AI selectively remove weeds without harming crops.
- **Genetically Modified (GM) Crops**: Herbicide-tolerant crops, such as glyphosate-resistant soybean, allow for more effective weed control with reduced tillage.

Challenges and Future Directions

Despite its benefits, IWM faces challenges such as labor costs, knowledge gaps, and resistance management. Future research should focus on:

- **Developing New Biocontrol Agents**: Expanding the use of natural enemies for weed suppression.
- **Enhancing Herbicide Stewardship**: Educating farmers on responsible herbicide use to prevent resistance.
- **Integrating Digital Technologies**: Leveraging AI and IoT for real-time weed monitoring and management.

IWM represents a holistic and sustainable approach to weed control, ensuring long-term agricultural productivity while minimizing environmental impact.

H. Best Practices for Herbicide Application

Effective herbicide application is essential for achieving optimal weed control while minimizing environmental impact and herbicide resistance. Proper application techniques ensure that herbicides reach the target weeds in the right concentration, reducing wastage and improving efficacy. This section outlines the best practices for herbicide application based on scientific principles and field research.

1. Selecting the Right Herbicide

Choosing the appropriate herbicide is the first step in effective weed management. The selection should be based on:

- **Weed Species**: Identify the dominant weed species and their growth stages. Use selective herbicides for specific weeds and non-selective herbicides for broad-spectrum control.
- **Mode of Action (MOA)**: Rotate herbicides with different MOAs to prevent herbicide resistance. The Herbicide Resistance Action Committee (HRAC) classification can guide MOA selection.
- **Soil and Environmental Conditions**: Consider soil texture, organic matter, and moisture levels, as they influence herbicide efficacy and persistence.

2. Proper Timing of Application

The timing of herbicide application significantly affects its effectiveness. Key considerations include:

Pre-emergence vs. Post-emergence

- *Pre-emergence herbicides* should be applied before weed germination and require soil moisture for activation.
- *Post-emergence herbicides* should be applied when weeds are actively growing and at their most susceptible stage.

- **Weather Conditions**: Avoid spraying during high winds, extreme temperatures, or before rainfall to prevent drift, volatilization, and runoff.
- **Growth Stage of Weeds**: Younger weeds (2-4 leaf stage) are more susceptible to herbicides than mature weeds.

3. Application Equipment and Calibration

Proper equipment selection and calibration ensure uniform herbicide distribution and prevent over- or under-application.

- **Sprayer Type**: Choose between boom sprayers, backpack sprayers, or aerial application based on field size and weed density.

Nozzle Selection

- Use flat-fan nozzles for broadcast applications.
- Use cone nozzles for directed sprays.
- Low-drift nozzles reduce off-target movement.
- **Sprayer Calibration**: Regularly calibrate sprayers to ensure correct application rates. Calibration involves checking nozzle output, spray pressure, and travel speed.

4. Mixing and Handling Herbicides

Proper handling and mixing of herbicides are critical for safety and effectiveness. Follow these guidelines:

- **Read the Label**: Always follow manufacturer recommendations for mixing rates, tank mixtures, and safety precautions.
- **Mixing Order**: Use the *WALES* method for tank mixing:
    1. **W** – Wettable powders
    2. **A** – Agitate
    3. **L** – Liquid flowables
    4. **E** – Emulsifiable concentrates
    5. **S** – Surfactants and adjuvants
- **Use of Adjuvants**: Add surfactants, crop oils, or drift-reducing agents as recommended to enhance herbicide absorption and reduce losses.

5. Application Techniques to Minimize Drift and Runoff

Herbicide drift and runoff can cause environmental contamination and reduce application efficiency. Best practices include:

- **Use of Drift-Reducing Nozzles**: Low-drift nozzles produce larger droplets, reducing wind drift.

- **Adjusting Boom Height**: Keep the boom at the recommended height (typically 50 cm above the target) to minimize drift.
- **Spraying Under Optimal Conditions**: Apply herbicides when wind speeds are between 3-10 km/h and avoid spraying during temperature inversions.
- **Buffer Zones**: Maintain buffer zones near water bodies and sensitive crops to prevent contamination.

6. Herbicide Resistance Management

Repeated use of the same herbicide can lead to resistance in weed populations. To prevent resistance:

- **Rotate Herbicides**: Use herbicides with different MOAs in successive applications.
- **Tank Mixing**: Combine herbicides with different MOAs to reduce selection pressure on weeds.
- **Integrated Weed Management (IWM)**: Combine chemical control with cultural, mechanical, and biological methods for sustainable weed control.

7. Post-Application Practices

After herbicide application, follow these best practices to ensure safety and effectiveness:

- **Cleaning Equipment**: Thoroughly clean sprayers to prevent cross-contamination between fields.
- **Monitoring Weed Control**: Assess herbicide efficacy 7-14 days after application and take corrective actions if necessary.
- **Record Keeping**: Maintain records of herbicide applications, including date, rate, weather conditions, and weed response, to improve future management decisions.

I. Future Trends and Research Directions in Herbicide Resistance

Herbicide resistance is a growing challenge in modern weed management, threatening global agricultural productivity. As resistant weed populations continue to expand, researchers and industry professionals are actively exploring new strategies to mitigate resistance and develop sustainable weed control solutions. This section highlights emerging trends and research directions in herbicide resistance management.

1. Advancements in Herbicide Discovery and Mode of Action Studies

- ☆ **New Herbicide Modes of Action (MOA):** The discovery of novel herbicide targets is crucial to overcoming resistance. Researchers are focusing on identifying new biochemical pathways in weeds that can be exploited for herbicidal action. Advances in molecular biology and genomics are accelerating this process (Duke, 2020).
- ☆ **Synergistic Herbicide Combinations:** Combining multiple herbicides with different MOAs can delay resistance development. Research is ongoing to identify effective synergistic mixtures that enhance weed control while minimizing selection pressure for resistance.

2. Precision Weed Management and Digital Technologies

- ☆ **Site-Specific Herbicide Application:** Precision agriculture technologies, such as GPS-guided sprayers and variable-rate application systems, enable targeted herbicide use, reducing overall chemical inputs and limiting resistance selection.
- ☆ **Artificial Intelligence (AI) and Machine Learning:** AI-driven weed detection systems, integrated with autonomous sprayers, can differentiate between crops and weeds, allowing precise herbicide application only where needed. This reduces herbicide overuse and delays resistance evolution.
- ☆ **Remote Sensing and Drones:** High-resolution imaging from drones and satellites helps monitor weed populations and detect early signs of herbicide resistance, allowing timely intervention.

3. Genetic and Biotechnological Approaches

- ☆ **Gene Editing for Herbicide Sensitivity Restoration:** CRISPR-Cas9 and other gene-editing technologies are being explored to modify resistant weed populations, potentially restoring their susceptibility to herbicides.
- ☆ **Herbicide-Tolerant Crops with Multiple Resistance Traits:** Advances in biotechnology are enabling the development of crops with stacked herbicide tolerance traits, allowing the use of multiple herbicides in rotation to manage resistant weed species effectively.
- ☆ **RNA Interference (RNAi) for Weed Control:** RNAi-based herbicides are an emerging area of research, targeting specific weed genes to disrupt their growth without affecting crops or beneficial organisms (Sammons *et al.*, 2021).

4. Integrated Weed Management (IWM) Strategies

- **Diversified Crop Rotations:** Rotating crops with different growth habits and competitive abilities can reduce weed seed banks and lower reliance on herbicides.
- **Cover Crops and Mulching:** Using cover crops and organic mulches suppresses weed emergence, reducing the need for chemical control.
- **Mechanical and Cultural Practices:** Combining mechanical weeding with optimized planting densities and row spacing enhances weed suppression while minimizing herbicide dependence.

5. Evolutionary and Ecological Studies on Resistance Mechanisms

- **Understanding Resistance Evolution:** Research on the genetic and physiological mechanisms of herbicide resistance helps predict future resistance trends and develop proactive management strategies.
- **Fitness Costs of Resistance:** Some resistant weed populations exhibit reduced fitness in the absence of herbicide pressure. Exploiting these fitness costs through strategic herbicide rotations and non-chemical control methods is an area of active research.

6. Policy and Regulatory Innovations

- **Stricter Herbicide Stewardship Programs:** Regulatory agencies are implementing guidelines to promote responsible herbicide use, including mandatory resistance management plans for new herbicide registrations.
- **Incentives for Sustainable Practices:** Governments and industry stakeholders are exploring financial incentives for farmers adopting integrated weed management approaches to reduce herbicide reliance.

7. Public and Farmer Awareness Initiatives

- **Educational Programs on Resistance Management:** Extension services and research institutions are developing training programs to educate farmers on best practices for herbicide resistance prevention.
- **Collaborative Research Networks:** Global collaborations between scientists, industry, and policymakers are essential for sharing knowledge and developing coordinated resistance management strategies.

Herbicide resistance management requires a multi-faceted approach, integrating scientific advancements, technological innovations, and sustainable agricultural practices. Ongoing research in these areas will be critical in ensuring long-term weed control efficacy and agricultural sustainability.

Chapter 14

Integration of Data Science in Weed Management

The integration of data science into weed management has revolutionized how agricultural professionals approach weed control. By leveraging advanced computational tools, machine learning algorithms, and big data analytics, farmers and researchers can now make more informed decisions to optimize weed management strategies.

Applications of Data Science in Weed Management

- ☆ **Weed Identification and Mapping**: High-resolution satellite imagery, drones, and ground-based sensors combined with machine learning algorithms enable precise weed identification and spatial mapping. For example, convolutional neural networks (CNNs) are widely used for image-based weed classification, distinguishing weeds from crops with high accuracy (Li *et al.*, 2022).
- ☆ **Predictive Weed Modeling**: Data science facilitates the development of predictive models for weed emergence and growth patterns. These models use historical weather data, soil conditions, and cropping patterns to forecast weed infestations, allowing for timely interventions.
- ☆ **Precision Herbicide Application**: Algorithms analyze weed density and distribution data to guide site-specific herbicide application. This reduces chemical usage, minimizes environmental impact, and lowers costs.

Emerging Technologies

- **Artificial Intelligence (AI)**: AI-powered systems like autonomous robots equipped with weed-detection sensors are increasingly used for mechanical weeding.
- **Internet of Things (IoT)**: IoT devices collect real-time field data, enabling dynamic weed management decisions.

Challenges and Future Directions

Despite its potential, integrating data science into weed management faces challenges such as high costs, data standardization issues, and the need for user-friendly tools. Addressing these barriers will further enhance adoption and effectiveness.

A. Big Data Analytics in Weed Control

The integration of big data analytics into agriculture has revolutionized various aspects of crop production, including weed management. By leveraging large datasets, advanced computational tools, and predictive modeling, big data analytics enables precise, efficient, and sustainable weed control strategies. This section explores the role of big data in weed management, its applications, and the tools and technologies that make it possible.

1. Overview of Big Data in Agriculture

Big data refers to the collection, analysis, and interpretation of vast amounts of structured and unstructured data. In agriculture, big data encompasses information from diverse sources, such as satellite imagery, sensors, drones, weather data, and farm machinery. These datasets are processed using advanced algorithms to generate actionable insights.

In weed control, big data analytics helps in identifying weed infestations, predicting weed growth patterns, and optimizing herbicide application. The ability to process real-time data and historical trends allows for site-specific and timely interventions, reducing costs and environmental impact.

2. Components of Big Data Analytics in Weed Control

Big data analytics in weed management relies on several key components:

- **Data Collection**: Data is gathered from various sources, including:
- **Remote Sensing**: Satellite and drone imagery provide high-resolution data on weed distribution.
- **IoT Sensors**: Field sensors measure soil moisture, temperature, and other parameters influencing weed growth.
- **Farm Machinery**: GPS-enabled equipment records spatial data during field operations.

- ☆ **Data Storage and Processing**: Cloud computing platforms store and process large datasets. Tools like Hadoop and Apache Spark enable the efficient handling of data.
- ☆ **Machine Learning and Artificial Intelligence (AI)**: Algorithms analyze data to identify patterns, classify weed species, and predict weed emergence.
- ☆ **Visualization Tools**: Dashboards and Geographic Information Systems (GIS) present data insights in user-friendly formats for decision-making.

3. Applications of Big Data Analytics in Weed Control

Weed Detection and Mapping

Big data analytics enables precise weed detection and mapping through the integration of remote sensing and machine learning. High-resolution satellite and drone imagery are analyzed to differentiate weeds from crops based on spectral signatures. For example:

- ☆ **Normalized Difference Vegetation Index (NDVI)**: This index helps identify areas with abnormal vegetation growth, often indicative of weed infestations.
- ☆ **Convolutional Neural Networks (CNNs)**: These AI models classify weed species by analyzing image data.

Predictive Weed Growth Models

Predictive models use historical and real-time data to forecast weed emergence and growth. Factors such as soil temperature, moisture, and crop planting dates are analyzed to predict weed pressure. This allows farmers to time their weed control measures effectively, reducing the need for blanket herbicide applications.

Precision Herbicide Application

Big data analytics supports precision agriculture by enabling site-specific herbicide application. Weed maps generated from remote sensing data guide variable-rate sprayers to target only infested areas. This reduces herbicide usage, minimizes crop damage, and lowers environmental risks.

Integrated Weed Management (IWM)

Big data facilitates the integration of cultural, mechanical, and chemical weed control methods. For instance, data on weed seedbank dynamics can inform crop rotation plans, while real-time monitoring can optimize the timing of mechanical weeding.

4. Tools and Technologies in Big Data Weed Management

Remote Sensing Technologies

- **Satellite Imagery**: Platforms like Sentinel-2 and Landsat provide multispectral data for large-scale weed monitoring.
- **Drones**: Unmanned aerial vehicles (UAVs) equipped with multispectral or hyperspectral cameras capture detailed images of weed infestations.

IoT and Field Sensors

- **Soil Sensors**: Measure parameters like moisture and temperature to predict weed emergence.
- **Weather Stations**: Provide localized weather data critical for modeling weed growth patterns.

Machine Learning Algorithms

- **Support Vector Machines (SVMs)**: Used for weed-crop classification in image data.
- **Random Forests**: Predict weed emergence based on environmental variables.

Decision Support Systems (DSS)

- DSS platforms integrate big data analytics with user-friendly interfaces, enabling farmers to make informed decisions. Examples include WeedScout and Agremo, which provide weed identification and management recommendations.

5. Benefits of Big Data Analytics in Weed Control

The adoption of big data analytics in weed management offers several advantages:

- **Increased Efficiency**: Precise weed detection and targeted interventions reduce labor and input costs.
- **Environmental Sustainability**: Optimized herbicide application minimizes chemical runoff and soil contamination.
- **Enhanced Crop Yields**: Effective weed control reduces competition for resources, improving crop productivity.
- **Data-Driven Decision Making**: Real-time insights enable proactive and informed weed management strategies.

6. Challenges and Limitations

Despite its potential, big data analytics in weed control faces certain challenges:

- ☆ **Data Quality and Availability**: Inconsistent or incomplete datasets can affect the accuracy of analytics.
- ☆ **High Initial Costs**: Implementing big data technologies requires significant investment in hardware, software, and training.
- ☆ **Technical Expertise**: Farmers may lack the skills to interpret complex data outputs.
- ☆ **Connectivity Issues**: Rural areas with limited internet access may struggle to adopt cloud-based solutions.

7. Case Studies and Real-World Applications

Case Study: Precision Weed Management in Cornfields

A study conducted in the United States used drone imagery and machine learning to map weed infestations in cornfields. The data guided variable-rate herbicide application, reducing chemical usage by 30 per cent without compromising weed control efficacy (Shirzadifar *et al.*, 2021).

Real-World Application: Smart Sprayers

Companies like John Deere have developed smart sprayers equipped with AI and big data analytics. These sprayers use real-time weed detection to apply herbicides only where needed, significantly reducing input costs and environmental impact.

8. Future Prospects

The future of big data analytics in weed control lies in the integration of emerging technologies such as:

- ☆ **Blockchain**: Ensures data transparency and traceability in weed management practices.
- ☆ **5G Connectivity**: Enhances data transmission speeds, enabling real-time analytics.
- ☆ **Autonomous Robots**: Combine big data with robotics for automated weed control.

B. Decision Support Systems (DSS) for Weed Management

Decision Support Systems (DSS) are advanced tools that integrate data, models, and user-friendly interfaces to assist agricultural professionals in making informed decisions. In the context of weed management, DSS plays a pivotal role in optimizing control strategies, reducing herbicide reliance, and improving overall crop productivity. This section delves into the principles, components, and applications of DSS in modern weed management, providing insights into its relevance in precision agriculture.

Principles of Decision Support Systems in Weed Management

DSS for weed management is built on the principles of integrating scientific knowledge, real-time data, and computational models to provide actionable insights. These systems are designed to:

- **Enhance Decision-Making:** By processing complex datasets, DSS simplifies decision-making for farmers and agronomists.
- **Promote Sustainability:** DSS encourages the adoption of integrated weed management (IWM) practices, reducing environmental impacts.
- **Optimize Resource Use:** By tailoring weed control measures to specific field conditions, DSS minimizes the overuse of herbicides and other inputs.

The foundation of DSS lies in its ability to analyze spatial and temporal variability in weed populations, crop growth, and environmental factors. This is achieved through the integration of technologies such as Geographic Information Systems (GIS), remote sensing, and predictive modeling.

Components of a Weed Management DSS

A robust DSS for weed management comprises several interrelated components:

1. Data Collection Systems

- **Field Surveys:** Manual or automated surveys to assess weed density, species composition, and distribution.
- **Remote Sensing:** Use of drones, satellites, or proximal sensors to capture high-resolution imagery of weed infestations.
- **IoT Sensors:** Internet of Things (IoT) devices for monitoring soil moisture, temperature, and other environmental parameters influencing weed growth.

2. Data Processing and Analysis

- **Geospatial Analysis:** GIS tools analyze spatial patterns of weed infestations to identify hotspots.
- **Machine Learning Models:** Algorithms predict weed emergence and growth based on historical and real-time data.
- **Crop-Weed Interaction Models:** These models simulate competition between crops and weeds to estimate yield losses.

3. Decision-Making Interface

- **User-Friendly Dashboards:** Interactive platforms displaying weed maps, control recommendations, and economic analyses.

- ☆ **Mobile Applications:** Apps enabling farmers to access DSS outputs in the field.
- ☆ **Advisory Systems:** Integration with extension services to provide expert guidance.

4. Knowledge Base

- ☆ **Scientific Databases:** Repositories of weed biology, herbicide efficacy, and best management practices.
- ☆ **Expert Systems:** Rule-based systems incorporating expert knowledge for specific weed management scenarios.

Applications of DSS in Weed Management

DSS has diverse applications in modern weed management, ranging from site-specific control to long-term planning. Key applications include:

1. Site-Specific Weed Management (SSWM)

DSS enables precision weed control by tailoring interventions to specific field zones. For example:

- ☆ **Weed Mapping:** High-resolution weed maps generated by drones or satellites guide variable-rate herbicide applications.
- ☆ **Selective Spraying:** Automated sprayers equipped with DSS apply herbicides only to weed-infested areas, reducing chemical use and costs.

2. Herbicide Resistance Management

DSS helps combat herbicide resistance by recommending diversified control strategies. Features include:

- ☆ **Resistance Risk Assessment:** Models predict the likelihood of resistance development based on herbicide use history and weed biology.
- ☆ **Rotation and Mixture Planning:** DSS suggests optimal herbicide rotations and tank mixtures to delay resistance.

3. Economic Threshold Analysis

DSS calculates the economic threshold for weed control, ensuring interventions are cost-effective. This involves:

- ☆ Estimating potential yield losses due to weed competition.
- ☆ Comparing the costs of control measures with expected benefits.

4. Integrated Weed Management (IWM)

DSS supports IWM by integrating cultural, mechanical, biological, and chemical control methods. For instance:

- ☆ **Cover Crop Selection:** DSS recommends suitable cover crops to suppress weed growth.
- ☆ **Tillage Optimization:** Tools analyze the impact of different tillage practices on weed seedbank dynamics.

5. Climate-Smart Weed Management

With climate change altering weed-crop interactions, DSS provides adaptive strategies. Examples include:

- ☆ Predicting shifts in weed species distribution under changing climatic conditions.
- ☆ Recommending drought-tolerant crops or altered planting dates to outcompete weeds.

Case Studies of DSS in Weed Management

1. WeedSOFT

WeedSOFT is a DSS developed in the United States to assist farmers in selecting effective weed control strategies. Key features include:

- ☆ A database of herbicide efficacy against various weed species.
- ☆ Economic analysis tools to evaluate the profitability of different control options.
- ☆ Integration with GIS for site-specific recommendations.

2. AgriWeed

AgriWeed, a European DSS, focuses on sustainable weed management. It incorporates:

- ☆ Predictive models for weed emergence based on soil temperature and moisture.
- ☆ Recommendations for non-chemical control methods, such as mechanical weeding.
- ☆ Tools for assessing the environmental impact of herbicide applications.

3. RIM (Resistance Integrated Management)

Developed in Australia, RIM is a DSS specifically designed for managing herbicide resistance in annual ryegrass. It provides:

- ☆ Long-term simulations of resistance evolution under different management scenarios.

- Strategies for integrating cultural practices, such as delayed sowing and crop rotation.

Challenges and Limitations of DSS in Weed Management

Despite their potential, DSS for weed management faces several challenges:

1. Data Availability and Quality

- Accurate weed identification and mapping require high-quality data, which may not always be accessible.
- Variability in data collection methods can affect DSS outputs.

2. Adoption Barriers

- High initial costs of DSS tools and associated technologies (*e.g.*, drones, sensors) may deter small-scale farmers.
- Limited technical expertise among users can hinder effective implementation.

3. Model Accuracy

- Predictive models rely on assumptions that may not account for all field-specific variables.
- Uncertainty in climate projections can affect long-term recommendations.

4. Integration with Existing Systems

- DSS must be compatible with existing farm management software and machinery.
- Lack of standardization in data formats can impede integration.

Future Prospects for DSS in Weed Management

The future of DSS in weed management lies in leveraging emerging technologies and interdisciplinary approaches. Promising developments include:

- **Artificial Intelligence (AI):** AI-powered DSS can improve weed identification accuracy and predictive modeling capabilities.
- **Big Data Analytics:** Integration of large-scale datasets from multiple sources (*e.g.*, weather stations, remote sensing) will enhance decision-making precision.
- **Blockchain Technology:** Blockchain can ensure data transparency and traceability, fostering trust among stakeholders.
- **Collaborative Platforms:** Cloud-based DSS will enable real-time data sharing and collaboration among farmers, researchers, and policymakers.

C. Real-Time Monitoring and Alerts

Effective weed management in modern agriculture increasingly relies on precision technologies that enable real-time monitoring and alerts. These systems utilize advanced tools such as remote sensing, Internet of Things (IoT) devices, artificial intelligence (AI), and geographic information systems (GIS) to provide timely and accurate information on weed infestations. By integrating these technologies, farmers and agricultural professionals can make informed decisions, reduce herbicide use, and improve crop productivity. This section explores the principles, tools, and applications of real-time monitoring and alerts in weed management.

Principles of Real-Time Monitoring and Alerts

Real-time monitoring and alert systems are designed to detect, analyze, and communicate information about weed presence and growth dynamics. These systems operate on the following principles:

- ☆ **Data Acquisition**: Sensors, drones, and satellites collect data on weed distribution, density, and species composition. This data is often captured in real-time or near-real-time.
- ☆ **Data Processing**: Advanced algorithms and AI models process the collected data to identify weed patterns, assess their impact on crops, and predict future growth.
- ☆ **Decision Support**: The processed information is delivered to farmers or agronomists through user-friendly platforms, enabling timely interventions.
- ☆ **Automation**: Some systems integrate automated responses, such as activating precision sprayers or issuing alerts to mobile devices.

By leveraging these principles, real-time monitoring systems provide actionable insights, reducing the reliance on traditional, labor-intensive scouting methods.

Key Technologies in Real-Time Weed Monitoring

Several cutting-edge technologies underpin real-time weed monitoring and alert systems. These include:

1. Remote Sensing

Remote sensing technologies, such as drones and satellites, play a crucial role in detecting and mapping weeds over large areas. They use multispectral and hyperspectral imaging to differentiate between crops and weeds based on their spectral signatures.

- ☆ **Multispectral Imaging**: Captures data in specific wavelengths, such as visible and near-infrared, to identify vegetation stress and weed presence.

- **Hyperspectral Imaging**: Provides detailed spectral information, enabling the identification of specific weed species.

For example, drones equipped with multispectral cameras can fly over fields, capturing high-resolution images that are processed to identify weed patches. Satellite-based platforms like Sentinel-2 provide broader coverage, making them suitable for large-scale monitoring (Zhang *et al.*, 2021).

2. Internet of Things (IoT)

IoT devices, such as soil sensors and weather stations, enhance real-time monitoring by providing additional context for weed growth. These devices collect data on:

- Soil moisture and temperature
- Light intensity
- Humidity and rainfall

By integrating IoT data with weed monitoring systems, farmers can predict weed emergence and growth patterns more accurately.

3. Artificial Intelligence (AI) and Machine Learning (ML)

AI and ML algorithms are essential for processing large datasets generated by remote sensing and IoT devices. These technologies enable:

- **Weed Identification**: AI models trained on image datasets can distinguish weeds from crops with high accuracy.
- **Predictive Analytics**: ML algorithms analyze historical and real-time data to forecast weed growth and recommend control measures.

For instance, convolutional neural networks (CNNs) are widely used in image-based weed detection, achieving classification accuracies of over 90 per cent in controlled studies (Kamilaris and Prenafeta-Boldú, 2018).

4. Geographic Information Systems (GIS)

GIS platforms integrate spatial data from various sources, allowing for the visualization and analysis of weed distribution. These systems support:

- Mapping weed hotspots
- Tracking weed spread over time
- Planning site-specific weed control measures

GIS-based tools are particularly valuable for precision agriculture, where targeted interventions are critical.

Applications of Real-Time Monitoring and Alerts

Real-time monitoring and alert systems have diverse applications in weed management, including:

1. Early Detection of Weed Infestations

Timely detection is critical for effective weed control. Real-time systems enable farmers to identify weed infestations at an early stage, preventing them from competing with crops for resources. For example:

- Drones can detect weeds in their early growth stages, enabling targeted herbicide application.
- IoT sensors can monitor soil conditions conducive to weed germination, triggering alerts for pre-emergence herbicide application.

2. Site-Specific Weed Management

Real-time data supports site-specific weed management (SSWM), which involves tailoring control measures to specific field conditions. This approach reduces herbicide use and minimizes environmental impact. Examples include:

- Precision sprayers that apply herbicides only to weed-infested areas based on drone or satellite imagery.
- Variable-rate technology (VRT) that adjusts herbicide dosage according to weed density.

3. Integrated Weed Management (IWM)

Real-time monitoring systems are integral to IWM strategies, which combine cultural, mechanical, biological, and chemical control methods. For instance:

- Remote sensing data can guide mechanical weeding operations by identifying weed hotspots.
- IoT devices can monitor the effectiveness of biological control agents, such as weed-eating insects.

4. Risk Assessment and Forecasting

By analyzing environmental and historical data, real-time systems can predict weed outbreaks and assess their potential impact on crop yields. This information helps farmers prioritize control measures and allocate resources efficiently.

Benefits of Real-Time Monitoring and Alerts

The adoption of real-time monitoring and alert systems offers several advantages:

- **Increased Efficiency**: Automated data collection and analysis save time and labor compared to traditional scouting methods.
- **Improved Accuracy**: Advanced imaging and AI technologies provide precise information on weed species and distribution.

- **Cost Savings**: Targeted interventions reduce herbicide use and associated costs.
- **Environmental Sustainability**: Minimizing herbicide application lowers the risk of chemical runoff and soil contamination.
- **Enhanced Decision-Making**: Timely alerts enable proactive weed management, reducing crop losses.

Challenges and Limitations

Despite their potential, real-time monitoring and alert systems face several challenges:

- **High Initial Costs**: The deployment of drones, sensors, and AI models requires significant investment, which may be prohibitive for small-scale farmers.
- **Data Integration**: Combining data from multiple sources, such as remote sensing and IoT devices, can be complex and requires specialized expertise.
- **Connectivity Issues**: Real-time systems rely on stable internet connectivity, which may be limited in remote agricultural areas.
- **Algorithm Limitations**: AI models may struggle to differentiate between weeds and crops in complex field conditions, leading to false positives or negatives.

Addressing these challenges will require continued research, technological advancements, and the development of affordable solutions for smallholder farmers.

Future Trends in Real-Time Weed Monitoring

The field of real-time weed monitoring is evolving rapidly, with several promising trends on the horizon:

- **Integration with Autonomous Machinery**: Combining real-time monitoring systems with autonomous tractors and sprayers will enable fully automated weed control.
- **Advances in AI and Robotics**: Improved AI models and robotic systems will enhance weed detection accuracy and enable real-time mechanical weeding.
- **Cloud-Based Platforms**: Cloud computing will facilitate the storage and analysis of large datasets, making real-time monitoring systems more accessible.

- **Crowdsourced Data**: Farmers and researchers can contribute data to shared platforms, improving the accuracy and scalability of monitoring systems.

These developments hold the potential to revolutionize weed management, making it more efficient, sustainable, and accessible.

Chapter 15

Regulatory and Policy Aspects

Effective weed management is not only a technical challenge but also a regulatory concern. Governments and international organizations play a critical role in shaping policies to ensure sustainable weed control while minimizing environmental and health risks. This chapter explores the regulatory frameworks and policies governing weed management practices.

1. Pesticide Registration and Approval

Herbicides, as a primary tool for weed control, must undergo rigorous testing and approval processes before being marketed. Regulatory agencies, such as the U.S. Environmental Protection Agency (EPA) and the European Food Safety Authority (EFSA), evaluate herbicides for efficacy, toxicity, and environmental impact.

These evaluations include:

- ☆ Acute and chronic toxicity studies.
- ☆ Residue analysis in crops and soil.
- ☆ Environmental fate assessments, such as leaching potential and persistence.

2. Maximum Residue Limits (MRLs)

MRLs are established to ensure herbicide residues in food products remain within safe limits. These thresholds are regularly updated based on scientific research and international trade agreements (Codex Alimentarius Commission, 2023).

3. Weed Risk Assessment (WRA)

Regulatory frameworks often include WRA protocols to prevent the introduction and spread of invasive weed species. This involves analyzing the potential ecological and economic impact of non-native plants.

4. Integrated Weed Management (IWM) Mandates

Many countries now promote Integrated Weed Management (IWM) through policies that encourage reduced reliance on chemical herbicides. For instance, the European Union's Sustainable Use Directive (2009/128/EC) mandates the adoption of IWM practices.

5. Compliance and Monitoring

Farmers and agribusinesses must comply with pesticide application guidelines, buffer zone requirements, and record-keeping mandates. Regulatory bodies conduct periodic inspections to ensure adherence.

6. International Trade and Phytosanitary Measures

Weed contamination in exported agricultural products is a significant concern. Phytosanitary measures under the International Plant Protection Convention (IPPC) aim to prevent the spread of invasive weeds through global trade.

A. Compliance with Modern Agricultural Policies

Modern agricultural policies play a pivotal role in shaping sustainable farming practices, including weed management strategies. These policies are designed to ensure environmental protection, enhance agricultural productivity, and promote the responsible use of resources. Compliance with these policies is not only a legal obligation but also a critical step toward achieving long-term agricultural sustainability. This section explores the key aspects of compliance with modern agricultural policies in the context of weed management, emphasizing regulatory frameworks, best practices, and the role of technology.

1. Regulatory Frameworks Governing Weed Management

Governments and international organizations have established regulatory frameworks to guide weed management practices. These frameworks aim to mitigate the environmental and economic impacts of weeds while ensuring the safety of humans, animals, and ecosystems.

Pesticide Registration and Approval

- ☆ Herbicides, a primary tool in weed management, must undergo rigorous testing and registration processes before they can be used. Regulatory bodies such as the U.S. Environmental Protection Agency (EPA) and the European Food Safety Authority (EFSA) evaluate herbicides for efficacy, environmental impact, and human health risks.

- For example, glyphosate, one of the most widely used herbicides, has been subjected to extensive scrutiny to ensure its safe application under specific guidelines (EPA, 2021).

Integrated Pest Management (IPM) Mandates

- Many agricultural policies require the adoption of Integrated Pest Management (IPM) principles, which emphasize a combination of cultural, mechanical, biological, and chemical methods to control weeds. IPM reduces reliance on herbicides and minimizes environmental risks.
- For instance, the European Union's Common Agricultural Policy (CAP) encourages farmers to implement IPM as part of sustainable farming practices (European Commission, 2022).

Invasive Weed Control Regulations

- Policies often target the control of invasive weed species that threaten native ecosystems and agricultural productivity. Compliance involves monitoring, reporting, and eradicating invasive species as mandated by local or national authorities.

2. Best Practices for Policy Compliance in Weed Management

Farmers and agricultural professionals must adopt best practices to align with modern agricultural policies. These practices not only ensure compliance but also improve the efficiency and sustainability of weed management efforts.

Record-Keeping and Documentation

- Maintaining accurate records of herbicide applications, including the type, quantity, and timing of use, is essential for compliance. These records help demonstrate adherence to legal requirements and facilitate audits by regulatory authorities.
- Digital tools, such as farm management software, can streamline record-keeping and ensure data accuracy.

Adherence to Label Instructions

- Herbicide labels provide critical information on application rates, target weeds, and safety precautions. Strict adherence to label instructions is a legal requirement and a fundamental aspect of compliance.
- For example, applying herbicides at rates higher than recommended can lead to environmental contamination and legal penalties.

Buffer Zones and Drift Management

- Policies often mandate the establishment of buffer zones to prevent herbicide drift into sensitive areas such as water bodies, residential zones, or organic farms. Farmers must use appropriate equipment and techniques, such as low-drift nozzles, to minimize off-target movement.

Training and Certification

- Many countries require farmers and applicators to undergo training and obtain certification before using certain herbicides. These programs enhance knowledge of safe handling, application techniques, and environmental stewardship.

3. Role of Technology in Ensuring Compliance

Advancements in agricultural technology have revolutionized weed management and made it easier for farmers to comply with modern policies. Precision agriculture tools, in particular, have enhanced the accuracy and efficiency of weed control measures.

- **Remote Sensing and GIS**: Remote sensing technologies, such as drones and satellite imagery, enable farmers to monitor weed infestations and assess the effectiveness of control measures. Geographic Information Systems (GIS) help in mapping weed distribution and planning targeted interventions.
- **Smart Sprayers**: Smart sprayers equipped with sensors and artificial intelligence can identify weeds in real-time and apply herbicides only where needed. This reduces chemical usage, minimizes environmental impact, and ensures compliance with policies promoting sustainable practices.
- **Digital Compliance Tools**: Mobile apps and software platforms provide farmers with real-time updates on regulatory requirements, weather conditions, and herbicide application guidelines. These tools simplify compliance by offering actionable insights and reminders.

4. Challenges in Policy Compliance

While compliance with modern agricultural policies is essential, it is not without challenges. Understanding these challenges can help stakeholders develop strategies to overcome them.

- **Complexity of Regulations**: The sheer volume and complexity of regulations can be overwhelming for farmers, especially smallholders with limited resources. Simplified guidelines and extension services can bridge this gap.
- **Cost Implications**: Implementing policy-compliant practices often involves additional costs, such as purchasing certified herbicides, adopting new technologies, or undergoing training. Governments and

organizations can support farmers through subsidies and financial incentives.

- **Knowledge Gaps**: Lack of awareness or understanding of regulatory requirements can lead to unintentional non-compliance. Continuous education and outreach programs are crucial to address this issue.

5. Global Examples of Policy Implementation

Several countries have successfully implemented agricultural policies that promote sustainable weed management. These examples highlight the importance of policy compliance in achieving broader agricultural goals.

- **Australia's Biosecurity Act (2015)**: This legislation emphasizes the control of invasive weeds and pests through coordinated efforts between farmers, governments, and communities. Compliance involves regular monitoring and reporting of invasive species.
- **India's Insecticides Act (1968)**: Although primarily focused on pesticides, this act regulates the use of herbicides to ensure safety and efficacy. Farmers must obtain licenses and follow prescribed guidelines for herbicide application.

B. Standards for Robotic and AI-Driven Weed Control Systems

The advent of robotic and AI-driven weed control systems has revolutionized modern agriculture by offering precision, efficiency, and sustainability. However, the development and deployment of these technologies require adherence to specific standards to ensure safety, reliability, and effectiveness. This section outlines the key standards and guidelines for robotic and AI-driven weed control systems, focusing on their design, operation, and integration into agricultural ecosystems.

1. Design Standards for Robotic Weed Control Systems

The design of robotic weed control systems must meet stringent standards to ensure functionality, durability, and safe operation in diverse agricultural environments.

Hardware Requirements

- **Durability**: Robots must be designed to withstand harsh environmental conditions such as extreme temperatures, dust, moisture, and UV radiation.
- **Precision Components**: High-resolution cameras, LiDAR sensors, and GPS modules are essential for accurate weed detection and navigation.
- **Energy Efficiency**: Systems must incorporate energy-efficient motors and batteries to enable extended field operations without frequent recharging.

Safety Mechanisms

- **Collision Avoidance**: Robots must include sensors and algorithms to detect and avoid obstacles, ensuring safety for humans, animals, and crops.
- **Emergency Stop Systems**: A fail-safe mechanism should be integrated to halt operations immediately in case of malfunction or emergency.

Modular Design

- Modular components allow for easy maintenance, upgrades, and customization based on specific crop and weed management needs.

In-text citation: For instance, the International Organization for Standardization (ISO) has developed standards such as ISO 18497, which outlines safety requirements for autonomous agricultural machinery (ISO, 2018).

2. AI Standards for Weed Detection and Decision-Making

AI algorithms form the backbone of robotic weed control systems, enabling them to identify and manage weeds with precision. Adhering to AI standards ensures accuracy, fairness, and transparency.

Data Quality and Training

- AI models must be trained on diverse datasets that include images of various weed species, growth stages, and environmental conditions.
- Data labeling should be accurate and consistent to minimize errors in weed identification.

Algorithm Transparency

- Developers must document the decision-making process of AI algorithms to ensure transparency and accountability.
- Explainable AI (XAI) techniques should be employed to make the system's decisions understandable to end-users.

Bias Mitigation

- AI systems must be tested for biases that could lead to incorrect weed identification or discrimination against specific crop types.
- Regular updates and retraining of models are necessary to adapt to changing weed populations and environmental conditions.

Performance Metrics

- Accuracy, precision, recall, and F1-score are critical metrics for evaluating the performance of AI-driven weed detection systems.

- Field validation is essential to compare AI predictions with ground truth data.

In-text citation: The European Commission's Ethics Guidelines for Trustworthy AI emphasize the importance of transparency, accountability, and fairness in AI systems (European Commission, 2019).

3. Operational Standards for Field Deployment

Once designed and developed, robotic and AI-driven weed control systems must adhere to operational standards to ensure effective and safe field performance.

Calibration and Testing

- Robots must be calibrated before deployment to ensure accurate weed detection and herbicide application.
- Pre-deployment testing should include simulations and small-scale field trials to identify and resolve potential issues.

Integration with Farm Management Systems

- Robots should be compatible with existing farm management software to enable seamless data exchange and decision-making.
- Real-time data sharing allows farmers to monitor robot performance and make informed decisions.

Safety Protocols

- Operators must be trained to handle robotic systems safely, including understanding emergency procedures and maintenance requirements.
- Warning systems should alert operators to potential hazards, such as system malfunctions or environmental risks.

Regulatory Compliance

- Robots must comply with local and international regulations governing the use of autonomous machinery and herbicides.
- Environmental impact assessments should be conducted to ensure that robotic systems do not harm non-target organisms or ecosystems.

4. Environmental and Sustainability Standards

Robotic and AI-driven weed control systems must align with sustainability goals to minimize their environmental footprint.

Precision Application of Herbicides

- Robots should be equipped with precision sprayers to apply herbicides only to targeted weeds, reducing chemical usage and runoff.

- Variable rate technology (VRT) can be employed to adjust herbicide application rates based on weed density and type.

Energy Efficiency

- Solar panels or other renewable energy sources can be integrated into robotic systems to reduce reliance on fossil fuels.
- Energy-efficient algorithms should be used to optimize robot movement and operation.

Biodiversity Protection

- Systems must be designed to distinguish between weeds and beneficial plants, such as cover crops or pollinator-friendly species.
- Non-chemical weed control methods, such as mechanical weeding or laser-based systems, should be prioritized to reduce environmental impact.

5. Interoperability and Standardization

Interoperability ensures that robotic weed control systems can work seamlessly with other agricultural technologies and equipment.

Communication Protocols

- Robots must support standardized communication protocols, such as ISO 11783 (commonly known as ISOBUS), to enable data exchange with tractors, drones, and other machinery.
- Wireless communication standards, such as 5G or LoRaWAN, can facilitate real-time data transmission over large fields.

Data Formats

- Standardized data formats, such as GeoJSON or XML, should be used for recording and sharing weed maps, herbicide application data, and performance metrics.
- Open data standards promote collaboration and innovation by allowing researchers and developers to access and analyze system data.

Cross-Platform Compatibility

- Robotic systems should be compatible with multiple operating systems and software platforms to ensure widespread adoption and usability.

6. Ethical and Social Considerations

Ethical and social standards are crucial for the responsible development and deployment of robotic and AI-driven weed control systems.

Data Privacy

- Systems must comply with data protection regulations, such as the General Data Protection Regulation (GDPR), to safeguard farmers' data.
- Anonymization techniques should be employed to protect sensitive information.

Job Displacement Mitigation

- Developers and policymakers must address concerns about job displacement by providing training and support for workers transitioning to roles involving robotic systems.
- Collaborative robots (cobots) can be designed to assist rather than replace human workers.

Public Perception and Acceptance

- Transparent communication about the benefits and limitations of robotic weed control systems can help build trust among farmers and the public.
- Pilot programs and demonstrations can showcase the technology's potential and address concerns.

C. Global Trends in Bioherbicide Approvals and Usage

The increasing demand for sustainable agricultural practices has propelled bioherbicides into the spotlight as a promising alternative to synthetic chemical herbicides. Bioherbicides, derived from natural organisms such as bacteria, fungi, and plant extracts, offer environmentally friendly weed management solutions. This section explores the global trends in bioherbicide approvals and usage, highlighting regulatory frameworks, market dynamics, and adoption patterns in various regions.

1. Regulatory Frameworks Governing Bioherbicide Approvals

The approval process for bioherbicides varies significantly across regions, reflecting differences in regulatory priorities and agricultural policies. Regulatory agencies assess bioherbicides for efficacy, safety, and environmental impact before granting approval for commercial use.

- **United States**: The Environmental Protection Agency (EPA) oversees the registration of bioherbicides under the Federal Insecticide, Fungicide, and Rodenticide Act (FIFRA). The EPA has streamlined the approval process for biopesticides, including bioherbicides, to encourage innovation and adoption. Notable examples include *Phoma macrostoma*, a fungal bioherbicide approved for broadleaf weed control in turfgrass.

- ☆ **European Union**: The European Food Safety Authority (EFSA) regulates bioherbicides under Regulation (EC) No 1107/2009. The EU emphasizes the precautionary principle, requiring extensive data on environmental safety. The approval process is often slower compared to the U.S., but recent efforts to promote biopesticides align with the EU Green Deal's goal of reducing chemical pesticide use by 50 per cent by 2030.
- ☆ **Asia-Pacific**: Countries like India, China, and Australia are witnessing growing interest in bioherbicides. Regulatory frameworks in these regions are evolving to accommodate biopesticides, with a focus on reducing reliance on chemical herbicides to address environmental and health concerns.

2. Market Dynamics and Growth Drivers

The global bioherbicide market has experienced steady growth, driven by increasing awareness of sustainable agriculture and stringent regulations on chemical herbicides. According to a report by MarketsandMarkets (2022), the bioherbicide market is projected to grow at a compound annual growth rate (CAGR) of 14.5 per cent from 2022 to 2027.

Key Growth Drivers

1. **Environmental Concerns**: Rising awareness of the adverse effects of chemical herbicides, such as groundwater contamination and biodiversity loss, has spurred demand for bioherbicides.
2. **Consumer Preferences**: The growing preference for organic and residue-free produce has encouraged farmers to adopt bioherbicides in integrated weed management systems.
3. **Government Initiatives**: Subsidies and incentives for biopesticides in regions like the EU and North America have accelerated the adoption of bioherbicides.
4. **Technological Advancements**: Innovations in microbial formulation and delivery systems have improved the efficacy and shelf life of bioherbicides, making them more competitive with chemical alternatives.

3. Regional Adoption Patterns

The adoption of bioherbicides varies across regions, influenced by factors such as agricultural practices, crop types, and regulatory environments.

North America

- ☆ North America leads in bioherbicide adoption, driven by robust research and development (R&D) initiatives and favorable regulatory policies. For instance, *Pseudomonas fluorescens* strain ACK55 is widely used in the U.S. for controlling invasive weeds like leafy spurge (*Euphorbia esula*).

- Organic farming practices, which account for a significant share of agricultural land in the U.S. and Canada, have further boosted bioherbicide usage.

Europe

- Europe has seen a gradual increase in bioherbicide adoption, particularly in countries like France, Germany, and the Netherlands. The EU's Farm to Fork Strategy, which promotes sustainable agriculture, has created a conducive environment for bioherbicide adoption.
- However, challenges such as high production costs and limited availability of bioherbicides have hindered widespread adoption.

Asia-Pacific

- The Asia-Pacific region is emerging as a significant market for bioherbicides, with countries like India and China investing in R&D to develop cost-effective solutions. For example, *Parthenium hysterophorus*, a noxious weed in India, is being targeted using fungal bioherbicides like *Alternaria alternata*.
- The region's diverse cropping systems and high weed pressure present substantial opportunities for bioherbicide adoption.

Latin America

- Latin America, particularly Brazil and Argentina, is witnessing growing interest in bioherbicides due to the expansion of organic farming and concerns over herbicide-resistant weeds. Bioherbicides like *Myrothecium verrucaria* are being explored for controlling invasive species in sugarcane and soybean fields.

4. Challenges and Opportunities

Despite the promising potential of bioherbicides, several challenges must be addressed to enhance their adoption and effectiveness.

Challenges

- **Limited Spectrum of Activity**: Most bioherbicides target specific weed species, limiting their applicability in diverse cropping systems.
- **Production and Storage**: The mass production of microbial bioherbicides requires specialized facilities, and their shelf life is often shorter than chemical herbicides.

- **Cost Competitiveness**: The high cost of bioherbicides compared to synthetic herbicides remains a significant barrier for resource-constrained farmers.

Opportunities

- **Integrated Weed Management (IWM)**: Bioherbicides can play a crucial role in IWM strategies, complementing cultural, mechanical, and chemical control methods.
- **Public-Private Partnerships**: Collaboration between governments, research institutions, and private companies can accelerate the development and commercialization of bioherbicides.
- **Education and Training**: Raising awareness among farmers about the benefits and proper application of bioherbicides can drive adoption.

5. Future Outlook

The future of bioherbicides lies in overcoming current limitations through advancements in biotechnology and formulation techniques. Gene editing tools like CRISPR-Cas9 offer opportunities to enhance the efficacy of microbial bioherbicides by improving their weed-specificity and environmental stability. Additionally, the integration of bioherbicides with precision agriculture technologies, such as drone-based application systems, can revolutionize weed management practices.

Chapter 16

Challenges and Future Directions

Modern weed management faces several challenges that demand innovative approaches and adaptive strategies. These challenges are shaped by evolving agricultural practices, environmental concerns, and socio-economic factors. Below are the key challenges and potential future directions in weed management:

Challenges in Modern Weed Management

1. Herbicide Resistance

- ✰ The widespread and repeated use of herbicides has led to the evolution of herbicide-resistant weed species. For example, glyphosate-resistant *Amaranthus palmeri* (Palmer amaranth) has become a significant concern in row crops (Heap, 2023).
- ✰ Resistance to multiple herbicide modes of action further complicates control efforts.

2. Environmental Concerns

- ✰ Over-reliance on chemical herbicides has raised concerns about soil health, water contamination, and non-target species impact.
- ✰ Regulatory restrictions on certain herbicides, such as paraquat and atrazine, are limiting available tools for weed control.

3. Labor Shortages in Manual Weeding

- ✰ Declining availability of labor for manual weeding in many regions has increased reliance on mechanized and chemical methods, which may not always be sustainable.

4. **Climate Change**

   ☆ Rising temperatures and altered precipitation patterns are influencing weed growth dynamics, leading to the emergence of new invasive species and extended growing seasons.

Future Directions in Weed Management

1. **Integrated Weed Management (IWM)**

   ☆ Combining cultural, mechanical, biological, and chemical methods to reduce reliance on herbicides. For instance, cover cropping and crop rotation can suppress weed populations naturally.

2. **Advancements in Precision Agriculture**

   ☆ Technologies such as drones, GPS-guided sprayers, and machine learning algorithms are enabling site-specific weed control, reducing herbicide use and costs.

3. **Biological Control Innovations**

   ☆ Research into bioherbicides, such as *Phytophthora palmivora* for controlling *Striga* spp., offers promising alternatives to synthetic chemicals (Chandler *et al.*, 2011).

4. **Gene Editing and Crop Breeding**

   ☆ CRISPR-Cas9 technology is being explored to develop herbicide-tolerant crops and weed-suppressive traits in plants.

5. **Policy and Education**

   ☆ Strengthening farmer education on sustainable practices and implementing policies that promote research and adoption of non-chemical weed control methods.

By addressing these challenges and leveraging emerging technologies, the future of weed management can align with sustainable agricultural goals.

A. Overcoming Barriers to Technology Adoption

The adoption of modern weed management technologies has the potential to revolutionize agricultural practices, improving efficiency, sustainability, and profitability. However, several barriers hinder the widespread acceptance and implementation of these technologies. Understanding and addressing these challenges is critical for ensuring that farmers and agricultural professionals can fully benefit from advancements in weed control methods.

1. Economic Barriers

One of the most significant obstacles to technology adoption is the cost associated with acquiring and implementing new tools and systems. Many farmers, particularly those in developing regions or small-scale operations, face financial constraints that limit their ability to invest in advanced weed management technologies.

- ☆ **High Initial Investment**: Technologies such as autonomous weeders, precision sprayers, and herbicide-resistant crop varieties often require substantial upfront capital.
- ☆ **Uncertain Return on Investment (ROI)**: Farmers may be hesitant to adopt new technologies without clear evidence of their economic benefits. Concerns about unpredictable weather, fluctuating crop prices, and market demand exacerbate this uncertainty.
- ☆ **Limited Access to Credit**: Inadequate access to affordable credit or financing options further restricts farmers' ability to invest in modern weed control solutions.

2. Knowledge and Skill Gaps

The successful implementation of advanced weed management methods often requires specialized knowledge and skills, which many farmers and agricultural workers may lack.

- ☆ **Lack of Training**: Technologies such as machine learning-based weed identification systems or drone-based herbicide application demand technical expertise. Without proper training programs, farmers may struggle to use these tools effectively.
- ☆ **Resistance to Change**: Traditional farming practices are deeply ingrained in many agricultural communities. Farmers may be reluctant to adopt unfamiliar technologies, particularly if they perceive them as complex or unreliable.
- ☆ **Information Gaps**: Limited access to reliable information about the benefits, functionality, and maintenance of new technologies can hinder adoption. Extension services and agricultural advisory systems may not always provide adequate support.

3. Infrastructure and Accessibility Challenges

The availability of infrastructure and resources plays a critical role in determining the feasibility of adopting modern weed management technologies.

- ☆ **Connectivity Issues**: Many advanced technologies, such as GPS-guided weed control systems or cloud-based data analytics platforms, rely on stable internet connectivity. Rural areas with poor network coverage face significant challenges in utilizing these tools.

- **Equipment Availability**: In some regions, the supply chain for advanced agricultural machinery and tools is underdeveloped, making it difficult for farmers to access the necessary equipment.
- **Maintenance and Support**: The lack of local service centers or technicians for repairing and maintaining advanced equipment can deter farmers from investing in these technologies.

4. Social and Cultural Barriers

Social and cultural factors also influence the adoption of modern weed management technologies.

- **Peer Influence**: Farmers often rely on the experiences and recommendations of their peers when deciding whether to adopt new technologies. If early adopters face challenges or express dissatisfaction, others may be discouraged from following suit.
- **Gender Inequities**: In many agricultural communities, women play a significant role in farming but may have less access to resources, training, and decision-making power. This can limit the adoption of technologies that require collaborative efforts or financial investment.
- **Perceived Risks**: Farmers may fear potential risks associated with new technologies, such as crop failure, environmental harm, or regulatory penalties.

5. Policy and Regulatory Constraints

Government policies and regulations can either facilitate or hinder the adoption of modern weed management technologies.

- **Lack of Incentives**: Inadequate subsidies, tax breaks, or other financial incentives for adopting sustainable weed control methods can slow down technology adoption.
- **Regulatory Hurdles**: Stringent approval processes for new herbicides, genetically modified crops, or autonomous machinery may delay their availability in the market.
- **Inconsistent Policies**: Variability in agricultural policies across regions or countries can create uncertainty for farmers and technology providers.

Strategies to Overcome Barriers

To address these barriers, a multi-faceted approach involving stakeholders across the agricultural sector is essential. Key strategies include:

1. **Financial Support**: Governments and financial institutions should provide subsidies, low-interest loans, and grants to help farmers invest in new technologies.

2. **Capacity Building**: Training programs and workshops should be organized to equip farmers with the knowledge and skills needed to use advanced weed management tools effectively.
3. **Infrastructure Development**: Investments in rural connectivity, supply chains, and service networks can improve access to modern technologies.
4. **Awareness Campaigns**: Extension services, farmer cooperatives, and industry stakeholders should work together to disseminate information about the benefits and practical applications of new technologies.
5. **Policy Reforms**: Governments should streamline regulatory processes, promote public-private partnerships, and create an enabling environment for innovation in agriculture.

B. Addressing Weed Resistance in Modern Systems

Weed resistance to herbicides is a growing challenge in modern agriculture, threatening crop productivity and sustainability. Resistance occurs when weed populations evolve mechanisms to survive herbicide applications that previously controlled them. The phenomenon is driven by repeated use of herbicides with the same mode of action, coupled with limited integration of non-chemical weed management strategies. This section explores the mechanisms of resistance, strategies to mitigate its development, and modern tools to address this pressing issue.

Mechanisms of Weed Resistance

Weed resistance to herbicides can arise through various mechanisms, which are broadly categorized into target-site resistance (TSR) and non-target-site resistance (NTSR):

Target-Site Resistance (TSR)

- TSR occurs when mutations in the target enzyme of a herbicide reduce its binding affinity, rendering the herbicide ineffective. For instance, resistance to acetolactate synthase (ALS)-inhibiting herbicides often results from point mutations in the ALS gene (Tranel and Wright, 2002).
- Example: Amaranthus species, such as Palmer amaranth (*Amaranthus palmeri*), have developed TSR to glyphosate due to mutations in the EPSPS (5-enolpyruvylshikimate-3-phosphate synthase) enzyme.

Non-Target-Site Resistance (NTSR)

- NTSR involves mechanisms that reduce herbicide efficacy without altering the target site. These include enhanced herbicide metabolism, reduced herbicide absorption, or sequestration.

- ☆ Example: Blackgrass (*Alopecurus myosuroides*) in Europe has developed NTSR to multiple herbicides through enhanced metabolic detoxification mediated by cytochrome P450 enzymes.

Understanding these mechanisms is critical for designing effective weed management strategies that prevent or delay resistance.

Factors Driving Weed Resistance

Several factors contribute to the rapid evolution of herbicide resistance in weed populations:

1. **Overreliance on Herbicides:** Continuous use of herbicides with the same mode of action exerts strong selection pressure, favoring resistant biotypes.
2. **Monoculture Practices:** Lack of crop rotation reduces diversity in weed control tactics, allowing resistant weeds to dominate.
3. **Insufficient Integrated Weed Management (IWM):** Limited adoption of cultural, mechanical, and biological weed control methods exacerbates resistance risks.
4. **Seed Bank Persistence:** Weed seeds can remain viable in the soil for years, serving as a reservoir for resistant populations.

Strategies to Mitigate Weed Resistance

To address weed resistance, a multi-faceted approach is essential. The following strategies are recommended for sustainable weed management:

1. Herbicide Rotation and Mixtures

- ☆ Rotate herbicides with different modes of action to reduce selection pressure on a single target site.
- ☆ Use tank mixtures of herbicides with complementary modes of action to target multiple weed pathways simultaneously.
- ☆ Example: Combining glyphosate (EPSPS inhibitor) with a pre-emergent herbicide like metribuzin (PSII inhibitor) can delay resistance in glyphosate-tolerant crops.

2. Integrated Weed Management (IWM)

- ☆ IWM combines chemical, cultural, mechanical, and biological methods to manage weeds holistically:
- ☆ **Cultural Practices:** Rotate crops, adjust planting dates, and use competitive crop varieties to suppress weed growth.
- ☆ **Mechanical Control:** Employ tillage, mowing, or inter-row cultivation to physically remove weeds.

- **Biological Control:** Introduce natural enemies, such as insects or pathogens, to target specific weed species.

3. Precision Agriculture Tools

- Modern technologies, such as GPS-guided sprayers and weed-sensing systems, enable site-specific herbicide applications, reducing overall herbicide use and selection pressure.
- Example: Optical weed detection systems, like WeedSeeker®, identify and spray only the weeds, minimizing herbicide exposure to non-target plants.

4. Cover Crops and Mulching

- Cover crops, such as rye or clover, suppress weed emergence by shading the soil and releasing allelopathic compounds.
- Mulching with organic or synthetic materials creates a physical barrier that inhibits weed germination.

5. Monitoring and Resistance Testing

- Regularly monitor weed populations for early signs of resistance and conduct resistance testing to identify effective herbicide options.
- Example: Molecular assays can detect specific resistance-conferring mutations, such as the Pro106 mutation in glyphosate-resistant weeds.

Emerging Tools in Weed Resistance Management

Advances in biotechnology and data science are providing novel solutions to combat weed resistance:

RNA Interference (RNAi)

- RNAi-based herbicides target specific genes in weeds, offering high specificity and reduced environmental impact.
- Example: Research is underway to develop RNAi sprays targeting EPSPS genes in glyphosate-resistant weeds.

Gene Editing

- CRISPR-Cas9 technology holds potential for developing crops with enhanced weed competitiveness or modifying weed genomes to restore herbicide sensitivity.

Artificial Intelligence (AI)

- AI-driven weed identification systems analyze field images to distinguish resistant weeds, enabling timely interventions.

Case Study: Glyphosate Resistance in Palmer Amaranth

Palmer amaranth (*Amaranthus palmeri*) is a notorious example of herbicide resistance. This weed has evolved resistance to multiple herbicide classes, including glyphosate, due to both target-site mutations and gene amplification. Farmers in the U.S. have successfully managed resistant populations by integrating cultural practices, such as narrow-row planting and cover cropping, with herbicide rotation (Norsworthy *et al.*, 2012).

C. Vision for Future Innovations in Weed Management

The field of weed management is undergoing a transformative phase, driven by advancements in technology, ecological awareness, and the need for sustainable agricultural practices. Future innovations aim to address the limitations of current methods while ensuring environmental safety, economic viability, and long-term agricultural productivity. Below, we explore key areas of innovation that are shaping the future of weed management.

1. Precision Weed Management

Precision agriculture is revolutionizing weed control by integrating advanced technologies such as GPS, drones, and artificial intelligence (AI). Future weed management systems are expected to rely heavily on site-specific weed control, which minimizes the overuse of herbicides and reduces costs.

- ✰ **AI-Powered Weed Identification**: Machine learning algorithms, coupled with high-resolution imaging, can distinguish between crops and weeds with remarkable accuracy. For example, AI-based systems like Blue River Technology's "See and Spray" use cameras and sensors to apply herbicides only to weeds, reducing chemical usage by up to 90 per cent (Bongiovanni and Lowenberg-DeBoer, 2004).
- ✰ **Drones and Remote Sensing**: Drones equipped with multispectral cameras can map weed infestations across large fields, enabling farmers to target specific areas for treatment. This approach enhances efficiency and reduces environmental impact.

2. Robotics and Automation

The integration of robotics in agriculture is paving the way for autonomous weed management systems. These technologies are particularly promising for labor-intensive crops and regions facing labor shortages.

- ✰ **Autonomous Weeding Robots**: Robotic systems like the "FarmWise Titan" use AI and mechanical tools to remove weeds without harming crops. These robots are designed to operate continuously, offering a sustainable alternative to manual weeding and herbicide application.
- ✰ **Laser-Based Weed Control**: Emerging technologies are exploring the use of lasers to target and destroy weeds at the cellular level. This

method eliminates the need for chemicals and reduces the risk of herbicide resistance.

3. Biotechnological Innovations

Biotechnology holds immense potential for developing weed-resistant crops and novel bioherbicides. These innovations aim to enhance crop competitiveness while minimizing reliance on synthetic chemicals.

- ✰ **Gene Editing**: CRISPR-Cas9 and other gene-editing tools are being used to develop crop varieties with enhanced resistance to specific weeds or herbicides. For instance, researchers are working on crops that can outcompete weeds through allelopathy–the natural release of biochemicals that inhibit weed growth (Hussain *et al.*, 2021).
- ✰ **Bioherbicides**: Microbial-based herbicides, derived from naturally occurring organisms, are gaining attention as eco-friendly alternatives to synthetic chemicals. Examples include *Phoma macrostoma*, a fungus that selectively inhibits broadleaf weeds without affecting grasses.

4. Integrated Weed Management (IWM) with Digital Tools

Integrated Weed Management (IWM) combines multiple control methods to achieve long-term weed suppression. Future IWM strategies will leverage digital tools to enhance decision-making and implementation.

- ✰ **Decision Support Systems (DSS)**: Advanced DSS platforms, powered by big data and predictive analytics, can provide real-time recommendations for weed control based on weather, soil conditions, and weed biology.
- ✰ **Mobile Applications**: Smartphone apps are being developed to assist farmers in identifying weeds, tracking infestations, and selecting appropriate control measures. These tools democratize access to expert knowledge and improve on-farm decision-making.

5. Sustainable and Ecological Approaches

The increasing focus on sustainability is driving the adoption of ecological weed management practices. These approaches prioritize soil health, biodiversity, and long-term resilience.

- ✰ **Cover Cropping**: Cover crops such as clover and rye are being used to suppress weed growth by outcompeting them for resources. Future innovations may include genetically enhanced cover crops with improved weed-suppressing traits.
- ✰ **Soil Microbiome Manipulation**: Research is exploring ways to manipulate soil microbial communities to create environments unfavorable for weed germination and growth. This approach aligns with regenerative agriculture principles.

6. Herbicide Resistance Management

The widespread issue of herbicide-resistant weeds necessitates innovative strategies to mitigate resistance development.

- **Stacked Herbicide Traits**: Crops engineered with multiple herbicide resistance traits can allow for the rotation of herbicides with different modes of action, delaying resistance.
- **Non-Chemical Alternatives**: Techniques such as electrocution and thermal weeding are being refined to provide effective alternatives to chemical control.

7. Policy and Education Innovations

Future weed management will also depend on supportive policies and farmer education to ensure the adoption of innovative practices.

- **Regulatory Support for New Technologies**: Governments and regulatory bodies must streamline the approval process for novel weed control technologies, including bioherbicides and robotics.
- **Farmer Training Programs**: Educational initiatives will be crucial for equipping farmers with the knowledge and skills to implement advanced weed management systems effectively.

D. Artificial Intelligence and Machine Learning in Weed Control

The integration of Artificial Intelligence (AI) and Machine Learning (ML) into agriculture has revolutionized weed management practices. These technologies provide precise, efficient, and sustainable solutions to weed control challenges, reducing the reliance on traditional methods such as manual labor and blanket herbicide applications. By leveraging AI and ML, farmers and researchers can identify, monitor, and manage weeds with unprecedented accuracy, ultimately improving crop yields and reducing environmental impact.

Applications of AI and ML in Weed Control

AI and ML technologies have been applied in various aspects of weed management, including weed identification, mapping, and targeted control. Below are some key applications:

1. Weed Identification and Classification

AI-powered systems use computer vision and deep learning algorithms to identify and classify weeds based on their morphological and spectral characteristics. These systems rely on large datasets of weed images to train models capable of distinguishing between weeds and crops.

- **Example:** Convolutional Neural Networks (CNNs), a type of deep learning model, are widely used for image recognition tasks in agriculture. CNNs can analyze images captured by drones, robots, or smartphones to detect specific weed species in real time.

- ☆ **Advantage:** This approach reduces the risk of misidentification, enabling precise weed management strategies.

2. Weed Mapping

ML algorithms are used to create spatial weed maps by analyzing data collected from sensors, drones, or satellites. These maps provide detailed information on weed density, distribution, and growth stages across a field.

- ☆ **Example:** Unmanned Aerial Vehicles (UAVs) equipped with multispectral or hyperspectral cameras capture high-resolution images of fields. ML models process these images to generate weed distribution maps.
- ☆ **Benefit:** Weed mapping enables site-specific weed management, minimizing herbicide use and operational costs.

3. Targeted Weed Control

AI and ML facilitate the development of precision weed control systems that target weeds selectively without harming crops. These systems often integrate robotics and automation technologies.

- ☆ **Example:** Autonomous weed control robots, such as the "See and Spray" technology, use AI-driven cameras to detect weeds and apply herbicides only where needed. This reduces herbicide consumption and prevents chemical resistance in weeds.
- ☆ **Emerging Trend:** Some systems employ ML algorithms to control mechanical weeding tools, such as robotic arms or rotary hoes, for non-chemical weed management.

Key Technologies Enabling AI and ML in Weed Control

Several advanced technologies support the application of AI and ML in weed management:

1. Sensors and Imaging Technologies

High-resolution cameras, multispectral sensors, and hyperspectral sensors are essential for capturing data on weed and crop characteristics. These sensors provide information on plant morphology, reflectance, and chlorophyll content, which are critical for weed identification.

2. Cloud Computing and Big Data

AI and ML models require extensive computational power and large datasets for training and operation. Cloud computing platforms enable the storage and processing of vast amounts of agricultural data, while big data analytics extract actionable insights.

3. Internet of Things (IoT)

IoT devices, such as soil moisture sensors and weather stations, provide real-time data that can be integrated into AI models for dynamic weed management. For instance, IoT-enabled systems can adjust weed control strategies based on environmental conditions.

Benefits of AI and ML in Weed Control

The adoption of AI and ML in weed management offers numerous advantages:

- ✰ **Precision:** AI systems can differentiate between crops and weeds with high accuracy, reducing collateral damage to crops.
- ✰ **Efficiency:** Automated weed control systems save time and labor, allowing farmers to manage larger areas with fewer resources.
- ✰ **Sustainability:** By minimizing herbicide use and promoting site-specific management, AI-driven solutions reduce environmental impact.
- ✰ **Cost-Effectiveness:** Although the initial investment in AI technologies may be high, the long-term savings in herbicide and labor costs make them economically viable.

Challenges and Limitations

Despite their potential, AI and ML technologies face several challenges in weed control:

- ✰ **Data Quality and Availability:** The accuracy of AI models depends on the quality and diversity of training datasets. Insufficient or biased data can lead to errors in weed identification.
- ✰ **High Initial Costs:** The deployment of AI-powered systems requires significant investment in hardware, software, and training.
- ✰ **Complexity of Weed-Crop Interactions:** Variability in weed and crop growth patterns, influenced by environmental factors, can complicate AI model predictions.

Future Directions

The future of AI and ML in weed control lies in the development of more robust, adaptive, and user-friendly systems. Innovations such as federated learning, which allows AI models to learn from distributed datasets without compromising data privacy, could enhance the scalability of these technologies. Additionally, integrating AI with other emerging technologies, such as blockchain for traceability and augmented reality for real-time weed scouting, holds promise for advancing weed management practices.

E. Big Data Analytics and Predictive Modeling

The integration of **big data analytics** and **predictive modeling** into modern weed management is revolutionizing the agricultural industry. These advanced tools enable farmers, researchers, and agricultural professionals to make data-driven decisions, optimize weed control strategies, and improve crop productivity. By leveraging vast datasets and predictive algorithms, weed management can now be more precise, cost-effective, and environmentally sustainable.

Role of Big Data in Weed Management

Big data refers to the collection, processing, and analysis of large and complex datasets that traditional methods cannot handle. In agriculture, big data encompasses information from various sources, including:

- **Remote Sensing Data**: Satellite imagery, drones, and sensors provide high-resolution data on weed distribution, crop health, and soil conditions.
- **Weather Data**: Historical and real-time weather data help predict weed growth patterns and herbicide efficacy.
- **Field Data**: Information from GPS-enabled equipment, yield monitors, and soil sensors contributes to understanding field variability.
- **Genomic Data**: Advances in weed genomics provide insights into herbicide resistance mechanisms and genetic diversity.

By analyzing these datasets, agricultural professionals can identify weed hotspots, predict weed emergence, and tailor management practices to specific field conditions.

Predictive Modeling in Weed Management

Predictive modeling involves using statistical and machine learning algorithms to forecast future weed behavior based on historical and real-time data. These models are essential for proactive decision-making in weed control. Key applications include:

1. **Weed Emergence Prediction**: Models like the **WeedCast** system use temperature, soil moisture, and weed biology data to predict the timing of weed emergence. This helps farmers schedule pre-emergence herbicide applications more effectively.
2. **Herbicide Resistance Risk Assessment**: Predictive models analyze weed populations and herbicide usage patterns to estimate the likelihood of resistance development. For example, the **Herbicide**

Resistance Action Committee (HRAC) employs such models to guide resistance management strategies.

3. **Yield Loss Estimation**: Models such as **CROPSYST** simulate the impact of weed competition on crop yield, enabling farmers to prioritize control measures in high-risk areas.

Technologies Enabling Big Data and Predictive Modeling

Several technologies facilitate the collection and analysis of data for weed management:

- ✰ **Artificial Intelligence (AI) and Machine Learning (ML)**: These technologies process vast datasets to identify patterns and make predictions. For instance, convolutional neural networks (CNNs) are used to classify weed species from drone imagery.
- ✰ **Geographic Information Systems (GIS)**: GIS tools map weed infestations and overlay them with other spatial data, such as soil type and crop rotation history.
- ✰ **Internet of Things (IoT)**: IoT devices, such as soil moisture sensors and weather stations, provide real-time data for predictive models.

Benefits of Big Data Analytics and Predictive Modeling

The adoption of big data and predictive modeling in weed management offers several advantages:

- ✰ **Precision Weed Control**: Targeted herbicide applications reduce chemical usage and environmental impact.
- ✰ **Cost Savings**: Optimized weed management minimizes input costs while maximizing yield.
- ✰ **Sustainability**: Data-driven decisions promote integrated weed management (IWM) practices, reducing reliance on herbicides and preserving biodiversity.

Challenges and Future Directions

Despite its potential, the adoption of big data analytics and predictive modeling faces challenges such as data integration, high costs, and the need for technical expertise. Future research should focus on developing user-friendly tools and enhancing data-sharing platforms to make these technologies accessible to all stakeholders.

F. Role of Citizen Science in Weed Management

Citizen science, the active involvement of non-professional volunteers in scientific research, has emerged as a transformative tool in modern weed management. By leveraging the collective efforts of farmers, students, and community members, citizen science enables large-scale data collection, enhances public awareness, and fosters innovative solutions to weed-related challenges. This participatory approach has gained prominence in recent years due to advancements in technology and the growing need for sustainable weed control strategies.

Key Contributions of Citizen Science in Weed Management

1. Large-Scale Weed Monitoring and Mapping

- ✰ Citizen science plays a critical role in monitoring weed populations across vast geographic areas. Volunteers can report weed infestations using mobile applications, GPS devices, or online platforms.
- ✰ For example, platforms like iNaturalist and EDDMapS (Early Detection and Distribution Mapping System) allow users to upload geotagged photos of weeds, which are then verified by experts. This data helps researchers track the spread of invasive species, identify hotspots, and prioritize management efforts.
- ✰ Such large-scale monitoring would be prohibitively expensive and time-consuming without public participation.

2. Early Detection of Invasive Weeds

- ✰ Early detection is crucial for preventing the establishment and spread of invasive weeds. Citizen scientists, often being the first to observe unusual plant growth in their local areas, can provide timely alerts to authorities.
- ✰ A notable example is the detection of *Parthenium hysterophorus* (a highly invasive weed) in new regions through citizen-reported data. Early intervention based on such reports can significantly reduce management costs and ecological damage.

3. Data for Research and Policy Development

- ✰ The data collected by citizen scientists is invaluable for research on weed ecology, distribution patterns, and the effectiveness of control measures.
- ✰ Policymakers also benefit from this data, as it provides evidence-based insights for designing region-specific weed management programs. For instance, citizen science data has been used to inform national strategies for controlling *Ambrosia artemisiifolia* (common ragweed) in Europe (Bonney *et al.*, 2014).

4. Public Awareness and Behavioral Change

- ✰ Engaging citizens in weed management fosters a sense of ownership and responsibility for local ecosystems.

- Educational campaigns linked to citizen science projects help raise awareness about the ecological and economic impacts of weeds. For example, initiatives like the "Weed Spotters Network" in Australia train volunteers to identify and report invasive weeds, thereby empowering communities to take proactive measures.

Technological Enablers of Citizen Science in Weed Management

1. Mobile Applications and Digital Platforms

- Mobile apps like "Weed ID" and "PlantNet" simplify weed identification and reporting for non-experts. These tools often include image recognition features, making them accessible to a broader audience.
- Cloud-based platforms facilitate real-time data sharing and analysis, enabling researchers to respond quickly to emerging weed threats.

2. Remote Sensing and Drones

- While traditionally used by professionals, drones and remote sensing technologies are increasingly being adopted in citizen science projects. Volunteers can operate drones to capture high-resolution images of infested areas, which are then analyzed by experts.

3. Social Media and Crowdsourcing

- Social media platforms serve as effective channels for recruiting participants, disseminating information, and crowdsourcing data. Hashtags like #WeedWatch or #InvasivePlants have been used to mobilize citizen scientists globally.

Challenges and Limitations

Despite its potential, citizen science in weed management faces challenges such as data quality, volunteer retention, and the need for expert validation. Addressing these issues requires robust training programs, user-friendly tools, and continuous engagement with participants.

Case Study: Successful Citizen Science Initiative

The "Great Lakes Early Detection Network" (GLEDN) in the United States exemplifies the success of citizen science in weed management. This initiative relies on volunteers to report invasive weeds using a mobile app. The data collected has been instrumental in controlling species like *Phragmites australis* (common reed) and *Garlic Mustard* in the Great Lakes region (Crall *et al.*, 2015).

G. Global Collaboration and Knowledge Sharing

In the realm of modern weed management, global collaboration and knowledge sharing play a pivotal role in addressing the challenges posed by invasive weeds, herbicide resistance, and sustainable agricultural practices. By fostering international partnerships and leveraging shared expertise, agricultural professionals can develop innovative, science-based solutions to combat weeds effectively.

Importance of Global Collaboration

1. Addressing Transboundary Weed Challenges

Weeds are not confined by geographical boundaries. Invasive species such as *Parthenium hysterophorus* and *Amaranthus palmeri* have spread across continents, causing significant agricultural losses. Global collaboration enables countries to exchange data on the biology, spread, and control of such species, ensuring timely interventions.

- ☆ For example, the International Weed Genomics Consortium (IWGC) facilitates the sharing of genomic data to better understand herbicide resistance mechanisms in weeds (Kleinman *et al.*, 2022).

2. Standardizing Best Practices

Collaborative efforts help in creating standardized protocols for weed management. Organizations like the Food and Agriculture Organization (FAO) and the European Weed Research Society (EWRS) work to disseminate best practices for integrated weed management (IWM), ensuring that farmers worldwide adopt sustainable and effective methods.

3. Capacity Building

Developing nations often lack the resources and expertise to tackle complex weed issues. Through global partnerships, knowledge transfer programs, and training workshops, researchers and practitioners in these regions can access advanced technologies and methodologies.

Knowledge Sharing Platforms

1. Digital Databases and Open Access Journals

Online platforms such as the Global Invasive Species Database (GISD) and Weed Science Society of America (WSSA) provide free access to research articles, case studies, and weed identification tools. These resources enable professionals to stay updated on the latest advancements in weed management.

2. International Conferences and Workshops

Events like the International Weed Science Congress (IWSC) and regional symposia bring together experts, researchers, and policymakers to exchange ideas and present findings. These forums encourage cross-disciplinary collaboration and foster innovation.

3. Collaborative Research Projects

Joint research initiatives, such as those funded by the Consultative Group on International Agricultural Research (CGIAR), promote the development of region-specific weed control strategies. For instance, CGIAR's work on rice systems in Asia has led to improved weed management practices tailored to local conditions.

Role of Technology in Knowledge Sharing

Advances in technology have revolutionized knowledge dissemination. Tools like remote sensing, geographic information systems (GIS), and artificial intelligence (AI) are increasingly being used to map weed infestations and predict their spread. Mobile applications, such as WeedID and Plantix, allow farmers to identify weeds and receive management recommendations in real-time. These technologies bridge the gap between research and field application, ensuring that knowledge reaches end-users efficiently.

Case Study: Herbicide Resistance Management

The Herbicide Resistance Action Committee (HRAC) exemplifies global collaboration in tackling herbicide resistance. By pooling data from member countries, HRAC develops guidelines for resistance management, ensuring that farmers worldwide adopt practices that delay the evolution of resistant weed populations (Heap, 2023).

References

Ascard, J. (1995). Thermal Weed Control by Flaming: Biological and Technical Aspects. *Weed Research*, **35**(6), 377–385. https://doi.org/10.1111/j.1365-3180.1995.tb01635.x

Beckie, H. J., and Harker, K. N. (2017). Our top 10 herbicide-resistant weed management practices. *Pest Management Science*, **73**(6), 1045-1052.

Blue River Technology. (2021). See and Spray Technology. Retrieved from https://www.bluerivertechnology.com

Blue River Technology. (2022). See and Spray Technology. Retrieved from https://www.bluerivertechnology.com

Blue River Technology. (2023). See and Spray Technology. Retrieved from https://www.bluerivertechnology.com

Bongiovanni, R., and Lowenberg-DeBoer, J. (2004). Precision Agriculture and Sustainability. *Precision Agriculture*, **5**(4), 359-387.

Bonney, R., Shirk, J. L., Phillips, T. B., Wiggins, A., Ballard, H. L., Miller-Rushing, A. J., and Parrish, J. K. (2014). Next steps for citizen science. *Science*, **343**(6178), 1436-1437.

Brookes, G., and Barfoot, P. (2020). "Environmental impacts of genetically modified (GM) crop use 1996–2018: Impacts on pesticide use and carbon emissions." *GM Crops and Food*, **11**(4), 215-241.

Busi, R., Vila-Aiub, M. M., and Powles, S. B. (2013). Genetic control of 2,4-D resistance in Raphanus raphanistrum. *Herbicide Resistance in Plants*, **61**(4), 419-425.

Carbon Robotics. (2022). "Laser Weeding Technology for Sustainable Farming." Retrieved from https://carbonrobotics.com

Chalker-Scott, L. (2007). Impact of mulches on landscape plants and the environment–A review. *Journal of Environmental Horticulture*, **25**(4), 239–249.

Chandler, D., Bailey, A. S., Tatchell, G. M., Davidson, G., Greaves, J., and Grant, W. P. (2011). The development, regulation, and use of biopesticides for integrated pest management. *Philosophical Transactions of the Royal Society B: Biological Sciences*, **366**(1573), 1987–1998.

Chauhan, B. S., and Mahajan, G. (2014). Recent advances in weed management. Springer.

Chauhan, B. S., *et al.* (2021). "Advances in Nano Sensors for Weed Detection." *Journal of Precision Agriculture*, **22**(3), 345-360.

Chauhan, B. S., Mahajan, G., Sardana, V., Timsina, J., and Jat, M. L. (2021). "Weed Management in Conservation Agriculture Systems." *Advances in Agronomy*, **172**, 133-174.

Chen, M., Zhang, X., and Wang, Y. (2022). Advances in CRISPR-Cas9 applications for herbicide resistance in crops. *Journal of Agricultural Biotechnology*, **15**(3), 45-59.

Christensen, S., Søgaard, H. T., Kudsk, P., and Jørgensen, R. N. (2022). Advances in Precision Weed Management: A Review. *Weed Research*, **62**(3), 145-158.

Christensen, S., Søgaard, H. T., Kudsk, P., Nørremark, M., Lund, I., Nadimi, E. S., and Jørgensen, R. (2021). Site-specific weed control technologies. *Weed Research*, **61**(2), 123–136. https://doi.org/10.1111/wre.12456

Codex Alimentarius Commission. (2023). Pesticide Residues in Food and Feed. Retrieved from www.fao.org/fao-who-codexalimentarius

Crall, A. W., Newman, G. J., Jarnevich, C. S., Stohlgren, T. J., Waller, D. M., and Graham, J. (2015). Improving and integrating data on invasive species collected by citizen scientists. *Biological Invasions*, **17**(5), 1417-1429.

De Oliveira, J. L., Campos, E. V. R., and Fraceto, L. F. (2020). Recent advances in nanoherbicides for weed control. *Journal of Agricultural and Food Chemistry*, **68**(36), 9839–9852.

De Oliveira, J. L., Campos, E. V. R., Bakshi, M., Abhilash, P. C., and Fraceto, L. F. (2021). Application of nanotechnology for the encapsulation of herbicides: An environmentally safer approach for weed control. *Journal of Agricultural and Food Chemistry*, **69**(13), 3781-3796.

Doudna, J. A., and Charpentier, E. (2014). The new frontier of genome engineering with CRISPR-Cas9. *Science*, **346**(6213), 1258096.

Duke, S. O. (2012). Why have no new herbicide modes of action appeared in recent years? *Pest Management Science*, **68**(4), 505–512. doi:10.1002/ps.2333

Duke, S. O. (2018). "The history and current status of glyphosate." *Pest Management Science*, **74**(5), 1027-1034.

Duke, S. O. (2020). "Natural Products-Based Herbicides: Their Role in Weed Management." *Pest Management Science*, **76**(1), 7-20. https://doi.org/10.1002/ps.5587

Duke, S. O. (2020). "Why have no new herbicide modes of action appeared in recent years?" *Pest Management Science*, **76**(4), 1381-1389.

Duke, S. O. (2023). "New Developments in Herbicide Resistance and Weed Management." *Weed Science*, **71**(2), 145-160.

Duke, S. O., and Powles, S. B. (2008). Glyphosate: A once-in-a-century herbicide. *Pest Management Science*, **64**(4), 319-325.

Duke, S. O., and Powles, S. B. (2023). Glyphosate-resistant weeds and crops: Impacts and management strategies. *Weed Science*, **71**(1), 1-12.

Environmental Protection Agency (EPA). (2021). Overview of Glyphosate Usage and Safety. Retrieved from https://www.epa.gov

European Commission. (2009). Directive 2009/128/EC on the Sustainable Use of Pesticides. Official Journal of the European Union.

European Commission. (2019). Ethics guidelines for trustworthy AI. Retrieved from https://ec.europa.eu

European Commission. (2022). Integrated Pest Management in the Common Agricultural Policy. Retrieved from https://ec.europa.eu

Fennimore, S. A., and Slaughter, D. C. (2021). "Robotics, Automation, and Artificial Intelligence for Weed Management." *Weed Science*, **69**(4), 564-572. https://doi.org/10.1017/wsc.2021.29

Fernández-Moreno, P. T., Bastida, F., and De Prado, R. (2017). Glyphosate resistance in Echinochloa colona: A case study of resistance evolution in non-GMO cropping systems. *Weed Science*, **65**(4), 468-477.

Food and Agriculture Organization (FAO). (2017). "The Future of Food and Agriculture: Trends and Challenges." Rome: FAO.

Gaines, T. A., Zhang, W., Wang, D., Bukun, B., Chisholm, S. T., Shaner, D. L. and Patzoldt, W. L. (2010). Gene amplification confers glyphosate resistance in *Amaranthus palmeri. Proceedings of the National Academy of Sciences*, **107**(3), 1029-1034. https://doi.org/10.1073/pnas.0910879107

Grillo, R., Pereira, A. E. S., Nishisaka, C. S., de Lima, R., Oehlke, K., Greiner, R., and Fraceto, L. F. (2014). Chitosan/tripolyphosphate nanoparticles loaded with paraquat herbicide: An environmentally safer alternative for weed control. *Journal of Hazardous Materials*, **278**, 163-171. https://doi.org/10.1016/j.jhazmat.2014.05.074

Heap, I. (2014). Herbicide resistance in weeds: A global threat to food security. *New Phytologist*, **198**(4), 957-969.

Heap, I. (2022). "Herbicide Resistance in Weeds." *Weed Science Society of America*.

Heap, I. (2023). International Survey of Herbicide Resistant Weeds. Retrieved from www.weedscience.org

Heap, I. (2024). "The International Survey of Herbicide Resistant Weeds." *Weed Science Society of America*. Retrieved from www.weedscience.org.

Huang, Y., Thomson, S. J., and Reddy, K. N. (2022). "Integration of UAVs and AI for Precision Weed Management in Soybean Fields." *Precision Agriculture Journal*, **23**(4), 567-582.

Hussain, S., Siddique, T., Saleem, M., Arshad, M., and Khalid, A. (2021). Impact of Allelopathy on Weed Management and Crop Productivity. *Agronomy for Sustainable Development*, **41**(3), 1-16.

International Organization for Standardization (ISO). (2018). ISO 18497: Safety of highly automated agricultural machines. Retrieved from https://www.iso.org

Jampílek, J., and Králová, K. (2021). "Nanotechnology in Agriculture: Advances and Applications." *Nanomaterials*, **11**(4), 822.

Jha, P., Kumar, V., and Singh, S. (2021). "Advances in Precision Weed Management Using GIS and GPS Technologies." *Journal of Precision Agriculture*, **22**(3), 345–362.

John Deere. (2021). See and Spray Technology. Retrieved from https://www.deere.com

John Deere. (2022). See and Spray technology. Retrieved from https://www.deere.com

Kah, M., Beulke, S., Tiede, K., and Hofmann, T. (2013). Nanopesticides: State of knowledge, environmental fate, and exposure modeling. *Critical Reviews in Environmental Science and Technology*, **43**(16), 1823-1867.

Kah, M., Beulke, S., Tiede, K., and Hofmann, T. (2018). Nanopesticides: State of knowledge, environmental fate, and exposure modeling. *Critical Reviews in Environmental Science and Technology*, **48**(6), 417-468.

Kah, M., Beulke, S., Tiede, K., and Hofmann, T. (2018a). Nanopesticides: State of knowledge, environmental fate, and exposure modeling. *Critical Reviews in Environmental Science and Technology*, **48**(6), 467–499.

Kah, M., Kookana, R. S., Gogos, A., and Bucheli, T. D. (2018). A critical evaluation of nanopesticides: Benefits, risks, and future research needs. *Environmental Science: Nano*, **5**(3), 484–500. https://doi.org/10.1039/C7EN01112E

Kamilaris, A., and Prenafeta-Boldú, F. X. (2018). Deep learning in agriculture: A survey. *Computers and Electronics in Agriculture*, **147**, 70-90. https://doi.org/10.1016/j.compag.2018.02.016

Kasirajan, S., and Ngouajio, M. (2012). Polyethylene and biodegradable mulches for agricultural applications: A review. *Agronomy for Sustainable Development*, **32**(2), 501–529. https://doi.org/10.1007/s13593-011-0068-3

Kleinman, Z., *et al.* (2022). "Advancing Weed Genomics: International Collaboration and Data Sharing." *Weed Science Journal*, **70**(3), 345-360.

Kudsk, P., Mathiassen, S. K., and Moss, S. R. (2022). Integrated weed management: The role of precision agriculture and new technologies. *Weed Research*, **62**(1), 1-12.

Kumar, R., Singh, V., and Sharma, P. (2022). Advances in Biosensors for Weed Detection: A Review. *Journal of Agricultural Nanotechnology*, **14**(3), 45-58.

Kumar, S., *et al.* (2021). "Advances in Nano-Enabled Agriculture: Applications and Challenges." *Journal of Agricultural Science and Technology*, **23**(2), 45-58.

Lamb, D. W., Frazier, P., and Adams, P. (2021). "Remote sensing of weeds in agriculture: A review." *Precision Agriculture*, **22**(3), 345-367.

Li, J., Zhang, H., Wang, S., and Yu, J. (2022). CRISPR/Cas-mediated herbicide resistance in crops: Progress and prospects. *Plant Biotechnology Journal*, **20**(3), 345-356.

Li, J., *et al.* (2021). "RNA interference: A novel approach for weed management." *Weed Science*, **69**(3), 345–356.

Li, X., Zhang, Y., and Wang, J. (2022). "Deep learning for weed detection in precision agriculture: A review." *Computers and Electronics in Agriculture*, **198**, 107003.

Liakos, K. G., Busato, P., Moshou, D., Pearson, S., and Bochtis, D. (2018). Machine learning in agriculture: A review. *Sensors*, **18**(8), 2674. https://doi.org/10.3390/s18082674

Liebman, M., Mohler, C. L., and Staver, C. P. (2001). Ecological Management of Agricultural Weeds. Cambridge University Press.

Liebman, M., Mohler, C. L., and Staver, C. P. (2016). Ecological Management of Agricultural Weeds. Cambridge University Press.

Liu, Y., Chen, S., and Li, J. (2020). "Precision Agriculture Technologies for Weed Management." *Journal of Agricultural Engineering Research*, **45**(3), 123-135.

Losey, J. E., Rayor, L. S., and Carter, M. E. (1999). Transgenic pollen harms monarch larvae. *Nature*, **399**(6733), 214.

Lottes, P., Behley, J., Milioto, A., and Stachniss, C. (2017). Fully convolutional networks with sequential information for robust crop and weed detection in precision farming. *IEEE Robotics and Automation Letters*, **3**(4), 2870-2877. https://doi.org/10.1109/LRA.2018.2847027

Lottes, P., Khanna, R., Pfeifer, J., Siegwart, R., and Stachniss, C. (2017). UAV-based crop and weed classification for smart farming. *ISPRS Annals of the Photogrammetry, Remote Sensing and Spatial Information Sciences*, **IV-2/W3**, 151–158.

Macías, F. A., Molinillo, J. M. G., Varela, R. M., and Galindo, J. C. G. (2019). Allelopathy: A natural alternative for weed control. *Pest Management Science*, **75**(9), 2413–2427. https://doi.org/10.1002/ps.5406

MarketsandMarkets. (2022). Bioherbicides Market by Source, Application Mode, Formulation, Mode of Action, Crop Type, and Region - Global Forecast to 2027. Retrieved from https://www.marketsandmarkets.com

Milioto, A., Lottes, P., and Stachniss, C. (2018). Real-time semantic segmentation of crop and weed for precision agriculture robots leveraging RGB and NIR images. *Proceedings of the IEEE International Conference on Intelligent Robots and Systems (IROS)*, 1-8.

Milioto, A., Lottes, P., and Stachniss, C. (2018). Real-time semantic segmentation of crop and weed for precision agriculture robots leveraging RGB-D data. *Proceedings of the IEEE International Conference on Robotics and Automation (ICRA)*, 2229-2235.

Milioto, A., Lottes, P., and Stachniss, C. (2018). Real-time semantic segmentation of crop and weed for precision agriculture robots leveraging background knowledge in CNNs. *Robotics and Automation Letters*, **3**(4), 2870–2877. https://doi.org/10.1109/LRA.2018.2849499

Milioto, A., Lottes, P., and Stachniss, C. (2021). Real-time semantic segmentation of crop and weed for precision agriculture robots leveraging deep learning. *Frontiers in Plant Science*, **12**, 1-12. https://doi.org/10.3389/fpls.2021.635391

Mirsky, S. B., Ryan, M. R., and Teasdale, J. R. (2013). Cover crop-based weed management strategies. *Agronomy Journal*, **105**(3), 765-775.

Mirsky, S. B., Ryan, M. R., Curran, W. S., *et al.* (2017). "Weed suppression by cover crops in no-till systems: A meta-analysis." *Agronomy Journal*, **109**(1), 1-12.

Mirsky, S. B., Ryan, M. R., Teasdale, J. R., Curran, W. S., Reberg-Horton, S. C., Spargo, J. T., and Wells, M. S. (2011). Conservation tillage issues: Cover crop-based organic rotational no-till grain production in the mid-Atlantic region, USA. *Renewable Agriculture and Food Systems*, **26**(3), 193–202.

Mirsky, S. B., Ryan, M. R., Teasdale, J. R., Curran, W. S., Reberg-Horton, S. C., Spargo, J. T., and Wells, M. S. (2013). Overcoming weed management challenges in cover crop–based organic rotational no-till systems. *Weed Technology*, **27**(1), 193-203.

Mulla, D. J. (2013). Twenty-five years of remote sensing in precision agriculture: Key advances and remaining knowledge gaps. *Biosystems Engineering*, **114**(4), 358-371.

Neve, P., Busi, R., Renton, M., and Vila-Aiub, M. M. (2018). "Expanding the eco-evolutionary context of herbicide resistance research." *Pest Management Science*, **74**(11), 2350–2361.

Neve, P., Busi, R., Renton, M., and Vila-Aiub, M. M. (2018). Expanding the eco-evolutionary context of herbicide resistance research. *Pest Management Science*, **74**(11), 2353-2362.

Norsworthy, J. K., Ward, S. M., Shaw, D. R., Llewellyn, R. S., Nichols, R. L., Webster, T. M. and Barrett, M. (2012). Reducing the risks of herbicide resistance: Best management practices and recommendations. *Weed Science*, **60**(sp1), 31-62. https://doi.org/10.1614/WS-D-11-00155.1

Owen, M. D. K., and Zelaya, I. A. (2015). Herbicide-resistant crops and weed resistance to herbicides. *Pest Management Science*, **71**(3), 357-365.

Perez-Ortiz, M., Peña, J. M., Gutiérrez, P. A., Torres-Sánchez, J., Hervás-Martínez, C., and López-Granados, F. (2016). Selecting patterns and features for between- and within-crop-row weed mapping using UAV-imagery. *Expert Systems with Applications*, **47**, 85-94.

Perez-Ruiz, M., Slaughter, D. C., and Upadhyaya, S. K. (2020). "Autonomous robotic weed control systems: A review." *Computers and Electronics in Agriculture*, **169**, 105-113.

Peña, J. M., Torres-Sánchez, J., de Castro, A. I., Kelly, M., and López-Granados, F. (2013). Weed mapping in early-season maize fields using object-based analysis of unmanned aerial vehicle (UAV) images. *PLoS ONE*, **8**(10), e77151. https://doi.org/10.1371/journal.pone.0077151

Peña, J. M., Torres-Sánchez, J., de Castro, A. I., Kelly, M., and López-Granados, F. (2023). "Weed Mapping and Targeted Herbicide Application Using UAVs and Machine Learning." *Precision Agriculture*, **24**(1), 45-62. https://doi.org/10.1007/s11119-023-10012-8

Powles, S. B., and Yu, Q. (2010). "Evolution in Action: Plants Resistant to Herbicides." *Annual Review of Plant Biology*, **61**, 317-347.

Rico, C. M., Majumdar, S., Duarte-Gardea, M., Peralta-Videa, J. R., and Gardea-Torresdey, J. L. (2011). Interaction of nanoparticles with edible plants and their possible implications in the food chain. *Journal of Agricultural and Food Chemistry*, **59**(8), 3485–3498. https://doi.org/10.1021/jf104517j

Sammons, R. D., and Gaines, T. A. (2014). Glyphosate resistance: State of knowledge. *Pest Management Science*, **70**(9), 1367-1377.

Sammons, R. D., Gaines, T. A., and Duke, S. O. (2021). "RNAi-based herbicides: potential and challenges." *Weed Science*, **69**(4), 485-492.

Schmidt, S. M., *et al.* (2020). "Genome editing technologies in crop improvement." *Plant Biotechnology Journal*, **18**(5), 892–906.

Shamshiri, R. R., Kalantari, F., Ting, K. C., Thorp, K. R., Hameed, I. A., Weltzien, C. and Ehsani, R. (2018). Advances in greenhouse automation and controlled environment agriculture: A transition to plant factories and urban agriculture. *International Journal of Agricultural and Biological Engineering*, **11**(1), 1-22.

Shaner, D. L. (2014). Herbicide Handbook (10th ed.). Weed Science Society of America.

Shaner, D. L. (2014). Lessons Learned from the History of Herbicide Resistance. *Weed Science*, **62**(2), 427-431.

Shaner, D. L. (2014a). Lessons learned from the history of herbicide resistance. *Weed Science*, **62**(2), 427-431.

Shaner, D. L. (2020). "Herbicide Use in Precision Agriculture: Benefits and Challenges." *Weed Science*, **68**(4), 425–432.

Shaner, D. L., and Beckie, H. J. (2014). The future for weed control and technology. *Pest Management Science*, **70**(9), 1329-1339.

Shaner, D. L., and Beckie, H. J. (2021). Herbicide Resistance in Weeds and Crops. *Plant Physiology*, **185**(4), 1201–1213. https://doi.org/10.1104/pp.21.00123

Sharma, A., Singh, R., and Gupta, P. (2021). "Spatial Mapping of Invasive Weeds Using GIS and Remote Sensing." *Journal of Weed Science*, **69**(3), 245-256.

Sharma, A., Singh, S., and Kaur, G. (2021). Advances in controlled-release herbicide formulations: A review. *Journal of Agricultural and Food Chemistry*, **69**(12), 3456–3468. https://doi.org/10.1021/acs.jafc.0c07654

Shirzadifar, A., Moshiri, N., and Rahman, M. (2021). "Application of Drone-Based Remote Sensing for Precision Weed Management in Cornfields." *Journal of Precision Agriculture*, **22**(3), 245-258.

Shirzadifar, A., *et al.* (2022). "Machine Learning in Precision Agriculture: A Review of Weed Detection and Management." *Precision Agriculture Journal*, **23**(4), 567-589.

Singh, A., Gupta, R., and Verma, S. (2021). Nano Herbicides: A Sustainable Approach to Weed Management. *Nanotechnology in Agriculture*, **9**(2), 112-128.

Singh, P., Sharma, N., and Verma, R. (2022). "Advances in Nano Sensors for Precision Agriculture." *Nanotechnology in Agriculture and Food Science*, **45**(2), 123–140.

Slaughter, D. C., Giles, D. K., and Downey, D. (2008). Autonomous robotic weed control systems: A review. *Computers and Electronics in Agriculture*, **61**(1), 63-78. https://doi.org/10.1016/j.compag.2007.05.008

Smith, J., Brown, P., and Lee, R. (2022). "Development of VOC-Sensitive Nanosensors for Precision Weed Management." *Journal of Agricultural Nanotechnology*, **15**(3), 45-58.

Smith, J., Brown, R., and Lee, T. (2022). "Integrating Precision Agriculture with Agroecological Practices for Sustainable Weed Management." *Journal of Agricultural Science*, **180**(3), 245-259.

Smith, R. G., Mortensen, D. A., and Ryan, M. R. (2020). A review of autonomous weed control technologies. *Agricultural Systems*, **178**, 102763. https://doi.org/10.1016/j.agsy.2020.102763

Smith, R., Brown, P., and Taylor, J. (2021). "Emerging Technologies in Agricultural Remote Sensing." *Precision Agriculture Review*, **18**(2), 89-102.

Teasdale, J. R., Cavigelli, M. A., and Spargo, J. T. (2020). "Optimizing cover crop mixtures for weed suppression and soil health." *Agricultural Systems*, **178**, 102736.

Torres-Sánchez, J., Peña, J. M., de Castro, A. I., and López-Granados, F. (2015). Multi-temporal mapping of the vegetation fraction in early-season wheat fields using images from UAV. *Precision Agriculture*, **16**(1), 13-28.

Tranel, P. J., and Wright, T. R. (2002). Resistance of weeds to ALS-inhibiting herbicides: What have we learned? *Weed Science*, **50**(6), 700-712. https://doi.org/10.1614/0043-1745(2002)050[0700:ROWTAI]2.0.CO;2

VanGessel, M. J. (2001). Glyphosate-resistant horseweed (Conyza canadensis) in the Mid-Atlantic United States. *Weed Technology*, **15**(2), 447-451.

Vigneau, N., *et al.* (2021). "Effectiveness of Electro-Weeding in Sugar Beet Cultivation." *Journal of Agricultural Engineering*, **56**(4), 345-356.

Walia, U.S., Brar, L.S. and Dhaliwal, B.K. (1997). Resistance to isoproturon in *Phalaris minor* Retz. in Punjab. *Plant Protec. Quarterly*, **12**(3): 138-140.

Walsh, M. J., Newman, P., and Powles, S. B. (2013). Targeting weed seeds in-crop: A new weed control paradigm for global agriculture. *Weed Technology*, **27**(3), 431-436. https://doi.org/10.1614/WT-D-12-00181.1

Ward, S. M., Webster, T. M., and Steckel, L. E. (2013). Palmer amaranth (Amaranthus palmeri): A review. *Weed Technology*, **27**(1), 12-27.

Watson, A. K., and Wymore, A. (2020). Biological Control of Weeds: A World Overview. Cambridge University Press.

Weston, L. A., and Duke, S. O. (2003). Weed and crop allelopathy. *Critical Reviews in Plant Sciences*, **22**(3–4), 367–389.

Xiang, H., Zhang, Y., and Wang, J. (2021). "Hyperspectral Imaging for Precision Weed Management." *Journal of Precision Agriculture*, **22**(3), 345-360.

Zaller, J. G., *et al.* (2018). Glyphosate herbicide affects belowground interactions between earthworms and symbiotic mycorrhizal fungi in a model ecosystem. *Scientific Reports*, **8**(1), 1-11.

Zhang, C., and Kovacs, J. M. (2012). The application of small unmanned aerial systems for precision agriculture: A review. *Precision Agriculture*, **13**(6), 693–712. https://doi.org/10.1007/s11119-012-9274-5

Zhang, C., and Kovacs, J. M. (2021). The application of small unmanned aerial systems for precision agriculture: A review. *Precision Agriculture*, **22**(3), 713-739.

Zhang, C., and Kovacs, J. M. (2022). The application of unmanned aerial vehicles in precision agriculture: A review. *Precision Agriculture*, **23**(4), 1–25. https://doi.org/10.1007/s11119-021-09844-5

Zhang, C., Kovacs, J. M., and Walters, D. (2022). Precision Agriculture Technologies for Weed Management: A Review. *Precision Agriculture*, **23**(4), 789–812.

Zhang, C., Liu, X., and Wang, Y. (2022a). "AI-Driven Remote Sensing for Precision Weed Management." *Journal of Agricultural Technology*, **18**(3), 45-58.

Zhang, C., Walters, C., and Kovacs, J. M. (2022). "Applications of UAVs in Precision Agriculture: A Review." *Remote Sensing*, **14**(3), 1-20.

Zhang, C., Walters, D., and Kovacs, J. M. (2021). Applications of low altitude remote sensing in agriculture upon the advent of unmanned aerial vehicle technologies. *Advances in Agronomy*, **169**, 1-62. https://doi.org/10.1016/bs.agron.2021.07.001

Zhang, H., Zhang, J., Wei, P., *et al.*(2020). Genome editing in plants: Advances and applications. *Nature Reviews Molecular Cell Biology*, **21**(12), 661-677. https://doi.org/10.1038/s41580-020-00288-9

Zhang, L., Liu, H., and Zhao, J. (2021). RNA interference as a novel tool for weed management: Current progress and future prospects. *Weed Science*, **69**(4), 567-576.

Zhang, X., Huang, Y., and Zhao, C. (2022). Advances in Remote Sensing for Precision Agriculture: Applications in Weed Management. *Precision Agriculture Journal*, **23**(4), 567-584.

Zhang, X., Liu, Y., and Wang, J. (2022b). "Hyperspectral Imaging for Weed Detection in Precision Agriculture." *Remote Sensing*, **14**(15), 3654. https://doi.org/10.3390/rs14153654

Zhang, X., Wang, Y., and Li, J. (2022c). "Advances in Remote Sensing for Precision Agriculture." *Journal of Agricultural Science*, **15**(3), 45-60.

Index

D

E

F

G

H

I

J

K

L

M

N

O

P

Q

R

S

T

U

V

W